Methods in Molecular Biology

Series Editor
John M. Walker
School of Life and Medical Sciences
University of Hertfordshire
Hatfield, Herts., United Kingdom

For further volumes:
http://www.springer.com/series/7651

mRNA Processing

Methods and Protocols

Editor

Yongsheng Shi

Department of Microbiology and Molecular Genetics, School of Medicine, University of California, Irvine, CA, USA

Humana Press

Editor
Yongsheng Shi
Department of Microbiology and Molecular Genetics
School of Medicine
University of California
Irvine, CA, USA

ISSN 1064-3745 ISSN 1940-6029 (electronic)
Methods in Molecular Biology
ISBN 978-1-4939-7203-6 ISBN 978-1-4939-7204-3 (eBook)
DOI 10.1007/978-1-4939-7204-3

Library of Congress Control Number: 2017946167

Printed on acid-free paper

This Humana Press imprint is published by Springer Nature
The registered company is Springer Science+Business Media LLC
The registered company address is: 233 Spring Street, New York, NY 10013, U.S.A.

Preface

The year 2017 is the 40th anniversary of the landmark discovery of splicing by the laboratories of Richard J. Roberts and Phillip A. Sharp. At the time, it was unclear whether splicing was a rare event or a common phenomenon. With the completion of the Human Genome Project, we now know that splicing is a nearly universal mechanism for human mRNA maturation. More importantly, the great majority of human genes produce multiple mRNA isoforms through alternative mRNA processing, including alternative splicing and alternative polyadenylation. These alternative mRNA processing events are highly regulated in a tissue- and/or developmental stage-dependent manner. Aberrant regulation of mRNA processing has been causally linked to or implicated in many human diseases. Forty years of basic research into mRNA processing have culminated in another landmark event: the first antisense oligo that alters the splicing pattern of a human gene has been granted accelerated approval by the Food and Drug Administration for treating spinal muscular atrophy!

It is an exciting time to be studying mRNA processing. The vast majority of alternative mRNA processing events has remained an unexplored territory. The recent discovery of circular RNAs generated by "back-splicing" revealed a whole new type of mRNA processing. New technologies are being developed at a breathtaking rate that provides us with a broader and deeper view of the world of mRNA processing. For example, high-throughput sequencing methods provide an unbiased picture of mRNA processing of any biological samples under any condition. Metabolic labeling coupled with high-throughput sequencing can trace the life of all mRNAs, from synthesis, processing to degradation. For any specific mRNA processing event, high-throughput screens can be employed to sift through the entire genome to look for regulators. For any specific regulator that binds to RNA, we can map all of its RNA interactions in vivo in one experiment. Bioinformatic tools not only help reveal global trends and regulatory mechanisms but also predict mRNA processing alterations under different conditions. In addition to these tools for global analyses, the "old school" in vitro assays have also been developed for in-depth analyses of individual mRNA processing events. Innovative methods have been developed for purifying mRNA processing complexes that are suitable for functional and structural studies. Single particle methods enable us to follow individual mRNA processing reactions in real time. With such a wide range of available methodologies, it can be a daunting task for scientists in the mRNA processing field to pick and choose a method that is most suitable to their specific problems. In this volume, we have assembled a series of the most commonly used and state-of-the-art methodologies in the field of mRNA processing. It is our hope that these protocols will be helpful for researchers to explore the wonderful world of mRNA processing.

With all of these exciting new developments, I believe that the mRNA processing field is poised to enter an explosive growth period. If the past 40 years is any indication, the best is yet to come! Good luck with your experiments!

Irvine, CA, USA *Yongsheng Shi*

Contents

Contributors

EMAD BAHRAMI-SAMANI • *Department of Microbiology, Immunology, and Molecular Genetics, University of California, Los Angeles, Los Angeles, CA, USA*
KRISTIAN E. BAKER • *Center for RNA Science and Therapeutics, Case Western Reserve University School of Medicine, Cleveland, OH, USA*
STEVEN M. BLUE • *Department of Cellular and Molecular Medicine, University of California at San Diego, La Jolla, CA, USA; Stem Cell Program, University of California at San Diego, La Jolla, CA, USA; Institute for Genomic Medicine, University of California at San Diego, La Jolla, CA, USA*
JOERG E. BRAUN • *RNA Therapeutics Institute, Howard Hughes Medical Institute, and Department of Biochemistry & Molecular Pharmacology, University of Massachusetts Medical School, Worcester, MA, USA*
FRANCISCO CARRANZA • *Department of Microbiology and Molecular Genetics, University of California, Irvine, Irvine, CA, USA*
LUCA CARTEGNI • *Susan Lehman Cullman Laboratory for Cancer Research, Ernest Mario School of Pharmacy, Rutgers, The State University of New Jersey, Piscataway, NJ, USA*
DALEN CHAN • *Department of Pharmaceutical Sciences, University of California, Irvine, Irvine, CA, USA*
HONG CHENG • *Shanghai Key Laboratory of Molecular Andrology, State Key Laboratory of Molecular Biology, CAS Center for Excellence in Molecular Cell Science, Institute of Biochemistry and Cell Biology, Shanghai Institutes for Biological Sciences, Chinese Academy of Sciences, Shanghai, China*
ANDREY DAMIANOV • *Department of Microbiology, Immunology, and Molecular Genetics, The David Geffen School of Medicine, University of California, Los Angeles, Los Angeles, CA, USA*
ROBERT B. DARNELL • *Laboratory of Molecular Neuro-Oncology, Howard Hughes Medical Institute, The Rockefeller University, New York, NY, USA; New York Genome Center, New York, NY, USA*
CHAO FENG • *Department of Pharmaceutical Sciences, University of California, Irvine, Irvine, CA, USA*
ANGELA GARIBALDI • *Department of Microbiology and Molecular Genetics, University of California, Irvine, Irvine, CA, USA*
CHELSEA GELBOIN-BURKHART • *Department of Cellular and Molecular Medicine, University of California at San Diego, La Jolla, CA, USA; Stem Cell Program, University of California at San Diego, La Jolla, CA, USA; Institute for Genomic Medicine, University of California at San Diego, La Jolla, CA, USA,*
KLEMENS J. HERTEL • *Department of Microbiology and Molecular Genetics, University of California, Irvine, CA, USA*
KLEMENS J. HERTEL • *Department of Microbiology and Molecular Genetics, University of California, Irvine, Irvine, CA, USA*
C. KIONG HO • *Department of Infection Biology, Graduate School of Comprehensive Human Sciences, Faculty of Medicine, University of Tsukuba, Ibaraki, Japan*

HEATHER A. HUNDLEY • *Medical Sciences Program, Indiana University School of Medicine, Bloomington, IN, USA*
HUN-WAY HWANG • *Laboratory of Molecular Neuro-Oncology, Howard Hughes Medical Institute, The Rockefeller University, New York, NY, USA; Department of Pathology, University of Pittsburgh School of Medicine, Pittsburgh, PA, USA*
DONGMING LIANG • *Department of Biochemistry and Biophysics, University of Pennsylvania Perelman School of Medicine, Philadelphia, PA, USA*
ASHLEY L. LOUIE • *Department of Cellular and Molecular Medicine, University of California at San Diego, La Jolla, CA, USA; Stem Cell Program, University of California at San Diego, La Jolla, CA, USA; Institute for Genomic Medicine, University of California at San Diego, La Jolla, CA, USA*
MALIHEH MOVASSAT • *Department of Microbiology and Molecular Genetics, University of California, Irvine, CA, USA*
THAI B. NGUYEN • *Department of Cellular and Molecular Medicine, University of California at San Diego, La Jolla, CA USA; Stem Cell Program, University of California at San Diego, La Jolla, CA, USA; Institute for Genomic Medicine, University of California at San Diego, La Jolla, CA, USA*
ERIC L. VAN NOSTRAND • *Department of Cellular and Molecular Medicine, University of California at San Diego, La Jolla, CA, USA; Stem Cell ProgramUniversity of California at San DiegoLa JollaCAUSAInstitute for Genomic Medicine, University of California at San Diego, La Jolla, CA, USA*
EIMILE OAKES • *Genome, Cell and Developmental Biology Program, Indiana University, Bloomington, IN, USA*
JEONG EUN PARK • *Susan Lehman Cullman Laboratory for Cancer Research, Ernest Mario School of Pharmacy, Rutgers, The State University of New Jersey, Piscataway, NJ, USA*
NATOYA PEART • *Department of Medicine, University of Pennsylvania, Philadelphia, PA, USA*
GABRIEL A. PRATT • *Department of Cellular and Molecular Medicine, University of California at San Diego, La Jolla, CA, USA; Stem Cell Program, University of California at San Diego, La Jolla, CA, USA; Institute for Genomic Medicine, University of California at San Diego, La Jolla, CA, USA; Bioinformatics and Systems Biology Graduate Program, University of California at San Diego, La Jolla, CA, USA*
VICTOR SEREBROV • *RNA Therapeutics Institute, Howard Hughes Medical Institute, and Department of Biochemistry & Molecular Pharmacology, University of Massachusetts Medical School, Worcester, MA, USA*
HOSSEIN SHENASA • *Department of Microbiology and Molecular Genetics, University of California, Irvine, CA, USA*
YONGSHENG SHI • *Department of Microbiology and Molecular Genetics, School of Medicine, University of California, Irvine, Irvine, CA, USA*
MIN SHI • *Shanghai Key Laboratory of Molecular Andrology, State Key Laboratory of Molecular Biology, CAS Center for Excellence in Molecular Cell Science, Institute of Biochemistry and Cell Biology, Shanghai Institutes for Biological Sciences, Chinese Academy of Sciences, Shanghai, China*
JENNA E. SMITH • *Center for RNA Science and Therapeutics, Case Western Reserve University School of Medicine, Cleveland, OH, USA*

ROBERT C. SPITALE • *Department of Pharmaceutical Sciences, University of California, Irvine, Irvine, CA, USA; Department of Chemistry, University of California, Irvine, Irvine, CA, USA*

SHAYNA STEIN • *Department of Microbiology, Immunology, and Molecular Genetics, University of California, Los Angeles, Los Angeles, CA, USA*

DEIRDRE C. TATOMER • *Department of Biochemistry and Biophysics, University of Pennsylvania Perelman School of Medicine, Philadelphia, PA, USA*

BIN TIAN • *Department of Microbiology, Biochemistry and Molecular Genetics, Rutgers New Jersey Medical School, Rutgers Cancer Institute of New Jersey, Newark, NJ, USA*

PRANATHI VADLAMANI • *Biotechnology Professional Science Master's Program, Indiana University, Bloomington, IN, USA*

ERIC J. WAGNER • *Department of Biochemistry and Molecular Biology, University of Texas Medical Branch at Galveston, Galveston, TX, USA*

XIUYE WANG • *Department of Microbiology and Molecular Genetics, School of Medicine, University of California, Irvine, Irvine, CA, USA*

KE WANG • *Shanghai Key Laboratory of Molecular Andrology, State Key Laboratory of Molecular Biology, CAS Center for Excellence in Molecular Cell Science, Institute of Biochemistry and Cell Biology, Shanghai Institutes for Biological Sciences, Chinese Academy of Sciences, Shanghai, China*

RUTH WANG • *Department of Cellular and Molecular Medicine, University of California at San Diego, La Jolla, CA, USA; Stem Cell Program, University of California at San Diego, La Jolla, CA, USA; Institute for Genomic Medicine, University of California at San Diego, La Jolla, CA, USA*

JEREMY E. WILUSZ • *Department of Biochemistry and Biophysics, University of Pennsylvania Perelman School of Medicine, Philadelphia, PA, USA*

YI XING • *Department of Microbiology, Immunology, and Molecular Genetics, University of California, Los Angeles, Los Angeles, CA, USA*

GENE W. YEO • *Department of Cellular and Molecular Medicine, University of California at San Diego, La Jolla, CA, USA; Stem Cell Program, University of California at San Diego, La Jolla, CA, USA; Institute for Genomic Medicine, University of California at San DiegoLa JollaCAUSABioinformatics and Systems Biology Graduate Program, University of California at San Diego, La Jolla, CA, USA; Department of Physiology, Yong Loo Lin School of Medicine, National University of Singapore, Singapore, Singapore; Molecular Engineering Laboratory, A*STAR, Singapore, Singapore; Sanford Consortium for Regenerative Medicine, University of California at San Diego, La Jolla, CA, USA*

KYU-HYEON YEOM • *Department of Microbiology, Immunology, and Molecular Genetics, The David Geffen School of Medicine, University of California, Los Angeles, Los Angeles, CA, USA*

DINGHAI ZHENG • *Department of Microbiology, Biochemistry and Molecular Genetics, Rutgers New Jersey Medical School, Rutgers Cancer Institute of New Jersey, Newark, NJ, USA*

SIKA ZHENG • *Division of Biomedical Sciences, University of California Riverside, Riverside, CA, USA*

Chapter 1

Detection and Identification of Uncapped RNA by Ligation-Mediated Reverse Transcription Polymerase Chain Reaction

C. Kiong Ho

Abstract

The 5′-cap structure is an essential feature in eukaryotic mRNA required for mRNA stability and enhancement of translation. Ceratin transcripts are selectively silenced by decapping in the cytoplasm and later become translationally active again by acquiring the cap structure to regenerate translatable mRNAs. Identification of uncapped mRNA transcripts will reveal how gene expression is regulated by the mRNA recapping pathway. What follows is a sensitive method to detect and identify the uncapped mRNA from the cells. The technique consists of three parts: selective ligation of anchor RNA to the 5′-end of monophosphate RNA by double-strand RNA ligase, conversion of ligated RNA product into cDNA by reverse transcription, and amplification of a specific cDNA by polymerase chain reaction.

Key words RT-PCR, 5′-cap, Uncapped mRNA, mRNA recapping, RNA ligation, RNA processing, *Trypanosoma brucei*

1 Introduction

The 5′ end of eukaryotic mRNAs is modified by the addition of a methylguanosine cap (m7Gpp or cap 0) to protect mRNA from degradation and enhance protein synthesis (reviewed in [1]). Common belief is that this cap is removed by decapping enzyme cleaving m7Gpp from capped RNA and the remaining 5′-PO_4 RNA is thought to be rapidly degraded by a 5′-to-3′ exonuclease (Xrn1/Rat1) (reviewed in [2, 3]). Recent studies suggest that certain uncapped mRNAs are accumulated in a translationally quiescent state and may later become translationally active again by acquiring the cap structure, thereby shifting transcriptional regulation to the capping/decapping enzymes [4, 5]. However, uncapped transcripts cannot be efficiently detected because the majority of 5′-PO_4 mRNAs are either rapidly degraded by 5′-exonuclease or are present in low abundance.

Yongsheng Shi (ed.), *mRNA Processing: Methods and Protocols*, Methods in Molecular Biology, vol. 1648,
DOI 10.1007/978-1-4939-7204-3_1, © Springer Science+Business Media LLC 2017

Herein is described a sensitive method to detect uncapped mRNA from *Trypanosome brucei* by the ligation-mediated reverse transcription-polymerase chain reaction. Accumulation of uncapped transcripts was detected when mRNA recapping enzyme was depleted from *T. brucei* [6]. A schematic of ligation-mediated RT-PCR is shown in Fig. 1. The anchor RNA oligonucleotide is hybridized with a 48-mer splint DNA oligonucleotide that is designed to hybridize the 5′-end of the target RNA sequence, and is also complementary to the anchor RNA sequence at the 3′-end (Fig. 1). The mRNA that has undergone decapping possesses a 5′-PO_4 end, which can hybridize with the splint DNA to generate a duplex nick with 3′-OH and 5′-PO_4 that can be ligated by ATP-dependent RNA ligase. In contrast, RNA with 5′-cap or triphosphate-terminated ends cannot be ligated. In this protocol, the anchor RNA is selectively ligated to 5′-PO_4 RNA by dsRNA ligase (T4 Rnl2; type II ATP-dependent RNA ligase) directed by the splint DNA sequence. While type 1 ATP-dependent RNA ligase (T4 Rnl1) has been used to join the two strands of RNA using the splint DNA, the natural substrate for T4 Rnl1 is single-stranded tRNAs and thus, it is not optimal for joining a duplex nick [7]. In contrast, T4 Rnl2 can efficiently catalyze the ligation in a double-stranded nick in RNA or an RNA:DNA hybrid [8, 9, 11]. Another advantage of T4 Rnl2 is that it can also join 5′-PO_4 RNA with 2′-O RNA methylation on the first 5′-nucleoside (unpublished results) which is likely present on the decapped mRNA derived from the cap 1 structure. In the subsequent step, ligated RNA is converted to cDNA using oligo(dT) primer. The cDNA derived from ligated RNA can be detected by PCR using a primer specific to the cDNA 5′-anchor sequence. The PCR products can be directly cloned into plasmids for sequencing by TA cloning. This method can also be modified for other organisms or detect cleaved 5′-PO_4 RNA products, provided that 5′-sequence of the target RNA is known.

2 Materials

2.1 Oligonucleotides

5′-anchor RNA. 21-mer RNA 5′-r(CACUAUCGGAAUAAGGG CGACACG)-3′ (Fig.1a) (*see* **Note 1**).

Splint DNA. 5′-GTTCTAATAATAGCGTTAGTTCGTGTCGCC CTTATTCCGATAGTG-3′ (Fig. 1a).

5′-SL forward primer: DNA oligonucleotide with a sequence 7–32 of the spliced leader RNA: 5′-CGCTATTATTAGAACAGT TTCTGTAC-3′ for control reaction.

α-Tubulin reverse PCR primer: α-Tubulin (Tb927.1.2340) reverse primer 5′-GGAACAGCTGGCGGTACGTGCC-3′ for control reaction.

Oligo dT_{20}. DNA primer: 5′-TTTTTTTTTTTTTTTTTTTT-3′.

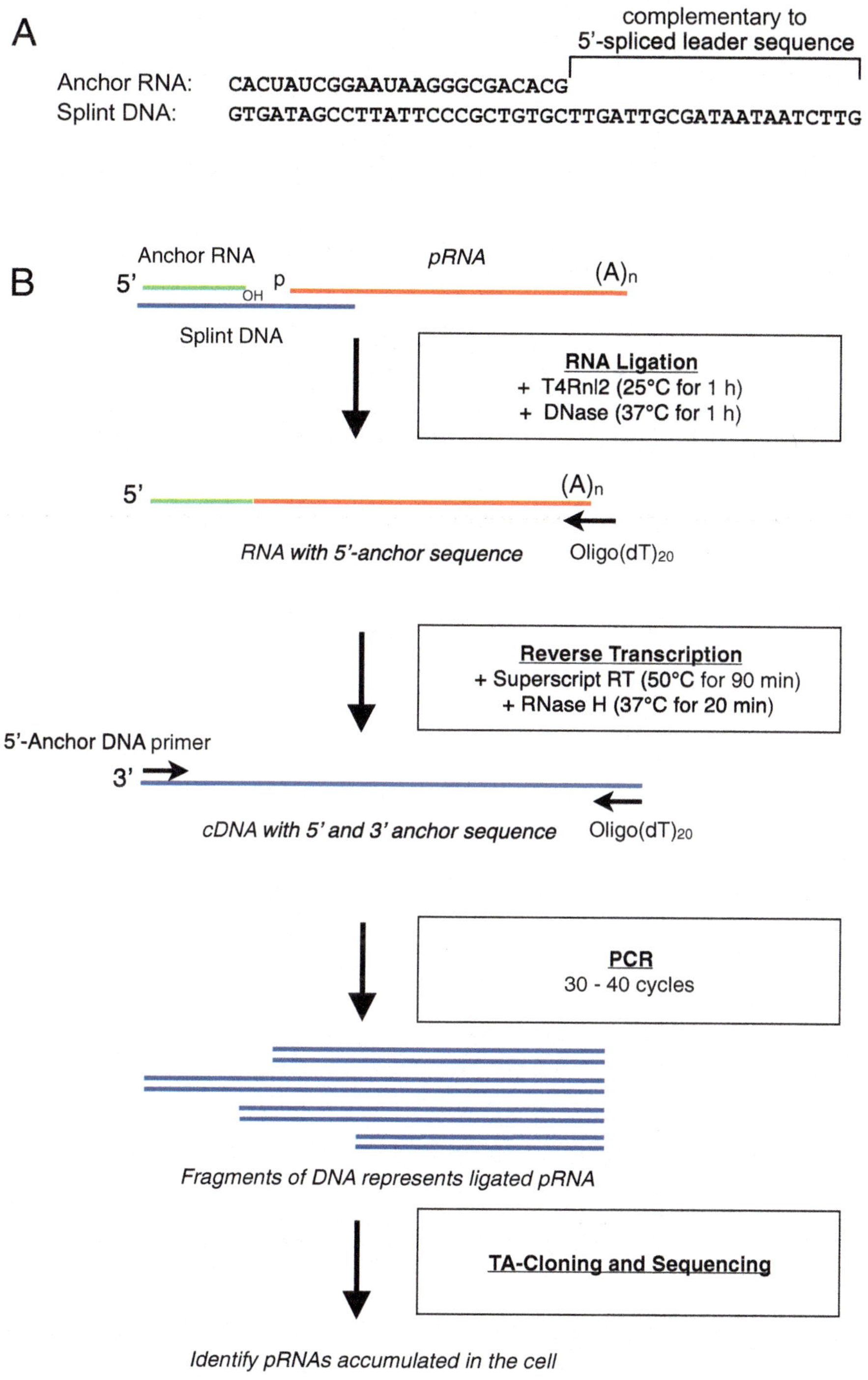

Fig. 1 Schematic of ligation-mediated RT-PCR. (**a**) The anchor RNA and splint DNA, whose 3′ half is complementary to the anchor RNA sequence and whose 5′ half is complementary to the SL RNA sequence are mixed with poly(A) RNA. (**b**) The ligation reaction was performed using T4 Rnl2, which preferentially joins nicks in duplex RNAs or RNA:DNA hybrids. Ligation-competent RNAs are detected by RT-PCR using DNA primers corresponding to the 5′-anchor and oligodT$_{20}$ for the PCR. Adapted from [6], Copyright (2015) National Academy of Sciences

2.2 Enzymes

- T4 RNA ligase 2 (T4 Rnl2) is commercially available from New England Biolabs. Alternatively, recombinant His-tagged T4 Rnl2 can be readily expressed and purified by affinity chromatography from bacteria [9]. Further purification of the His-tagged T4 Rnl2 through a sizing column or gradient centrifugation is recommended to remove trace amounts of RNase contamination in the final step of preparation.
- SuperScript™ Reverse Transcriptase III (Invitrogen/Thermo Fisher Scientific).
- RNase-free DNase 1 (10 U/μl).
- Ribonuclease Inhibitor (40 U/μl).
- TOPO® TA Cloning® Kit (Invitrogen/Thermo Fisher Scientific).

2.3 Buffers and Media

- TVM-1 medium supplemented with 20% FBS (1 l):

  For preparation, dissolve 1.16 g of L-Proline, 14.3 g of HEPES, 1 ml of Penicillin (100,000 U/ml)-Streptomycin (100,000 μg/ml) solution, in 650 ml Eagle's Minimum Essential Medium (EMEM). Adjust pH to 7.2 by 10 N NaOH and bring up the volume to 800 ml by EMEM.

  Add 200 ml of heat-inactivated fetal bovine serum and store at 4 °C.
- TE:

  10 mM Tris–HCl (pH 8).

  1 mM EDTA.
- Rnl2-ligation cocktail (5 μl per reaction):

  1.5 μl of 10 T4 Rnl2 Reaction Buffer.

  [0.5 M Tris–HCl (pH 7.5), 20 mM $MgCl_2$, 10 mM DTT, 4 mM ATP].

  3.5 μl of RNase-free H_2O.

  0.5 μl of T4 Rnl2 (0.25 μg/μl).
- DNase 10X Reaction Buffer:

  [400 mM Tris-HCl (pH 8), 100 mM$MgSO_4$ and 10 mM $CaCl_2$]
- RT cocktail (7 μl per reaction):

  4 μl of 5× First-Strand Buffer (supplied with Superscript™ III RT).

  [250 mM Tris–HCl (pH 8.3), 375 mM KCl, and 15 mM $MgCl_2$].

  1 μl of 0.1 M DTT.

  1 μl of RNase Inhibitor (40 units/μl).

  1 μl of SuperScript™ III RT(200 U/μl).

- Salt Solution for TOPO Cloning Reaction (provided in TOPO® TA Cloning® Kit):

  1.2 M NaCl.

  60 mM $MgCl_2$.

3 Methods

Follow full RNA hygiene measures against RNase contamination. Carry out all the procedures on ice unless otherwise specified.

3.1 RNA Preparation

1. Procyclic *T. brucei* 29.13 cells are cultivated in TVM-1 medium supplemented with 20% FBS (Subheading 2.3). Parasite density should be maintained between 1×10^6 and 6×10^6 cells/ml [10].
2. Total RNA will be purified from 0.5 to 2×10^8 cells using the Qiagen RNeasy Mini Kit according to the manufacturer's instructions.
3. Determine the concentration of RNA by spectroscopic measurement. Assess the integrity of total RNA by agarose gel electrophoresis stained. The absorbance ratio at 260/280 nm is an indicator of RNA quality (*see* **Note 2**).
4. Isolate the poly(A) RNA from approximately 150 μg of total RNA using a Takara Oligotex-dT30-Super mRNA Purification Kit according to the manufacturer's instructions.
5. Determine the concentration of the poly(A) RNA by spectroscopic measurement and store the sample in aliquots at −70 °C (*see* **Note 2**).

3.2 Splint Ligation by T4 Rnl2

1. Heat RNA sample (200 ng/μl) to 70 °C for 5 min and quick chill on ice.
2. Mix 50 pmol of anchor RNA and 50 pmol of splint DNA in a final volume of 5 μl in TE (Subheading 2.3) (*see* **Note 3**). For a control reaction, omit the splint DNA.
3. For each reaction, add 2 μl of RNA sample, 5 μl of oligonucleotide mixture, and 3 μl of 0.5 M NaCl. Incubate at 70 °C for 5 min, at 42 °C for 5 min, and then allow to cool to 23 °C in the water bath.
4. Add 5 μl of Rnl2-ligation cocktail (Subheading 2.3) to each reaction. Incubate for 30 min at room temperature.
5. Add 4 μl of DNase 10X Reaction Buffer (Subheading 2.3), 15 μl of water, and 1 μl of DNase (10 units/μl). Incubate for 1 h at 37 °C.
6. Bring up the sample volume to 200 μl with TE and extract with equal volume (200 μl) of phenol-chloroform at room temperature.

7. Transfer the aqueous phase to fresh tube. Precipitate the sample by the addition of 1 μl of glycogen, 20 μl (0.1 volume) of 3 M sodium acetate (pH 5.2), and 500 μl (2.5 volume) of ice-cold ethanol. Incubate on dry ice for 30 min or at −70 °C overnight.
8. Recover the RNA by centrifugation, wash with 500 μl of 70% ice-cold ethanol, and lyophilize the sample with speed vac. Resuspend in 20 μl of RNase-free water and store at −20 °C (*see* **Notes 4** and **5**).

3.3 cDNA Synthesis

1. Aliquot 2 μl of the splint-ligation reaction to a fresh 0.2 ml tube. Add 1 μl of 50 μM oligo(dT)$_{20}$ primer, 1 μl of 10 mM dNTP 9 μl of water. Incubate at 65 °C for 5 min and it is chilled on ice (*see* **Note 6**).
2. Add 7 μl of Superscript RT cocktail. Incubate at 50 °C for 90 min.
3. Terminate the reaction at 70 °C for 15 min, and incubate the sample with 1 μl of RNase H (40 U/μl) at 37 °C for 20 min to remove RNA complementary from the cDNA. Store at −20 °C.

3.4 PCR Amplification

1. Set up a PCR using 2 μl of cDNA product, in a reaction primed by 5′-anchor primer and oligo(dT)$_{20}$ primer (*see* **Note 6**).
2. My lab routinely performs 30–40 cycles of PCR with Taq DNA polymerase with the following parameters:

   Denaturation: 95 °C for 30 s.

   Annealing: 58–62 °C for 30 s.

   Polymerization: 72 °C for 120–150 s.
3. Analyze the PCR products by electrophoresis through a 0.8–1.2% agarose gel containing ethidium bromide (*see* **Notes 7** and **8**).
4. The efficiency of cDNA synthesis should be verified by PCR using an SL-DNA(7–32) forward primer and oligo(dT)$_{20}$ reserve primer (Fig. 2). For internal control experiments, PCR should also be carried out using SL-DNA(7–32) forward primer and gene-specific reverse primer, such as α-tubulin (*see* **Note 9**).

3.5 TOPO Cloning and Transformation

1. Set up a TOPO cloning reaction by incubating 2 μl of PCR product, 1 μl of Salt Solution (Subheading 2.3), 2 μl of water, and 1 μl of TOPO vector as described by manufacture's instruction.
2. Incubate for 5 min at room temperature and place the reaction on ice.
3. Transform 2 μl of TOPO cloning reaction into Mach1 or DH5α competent cells.

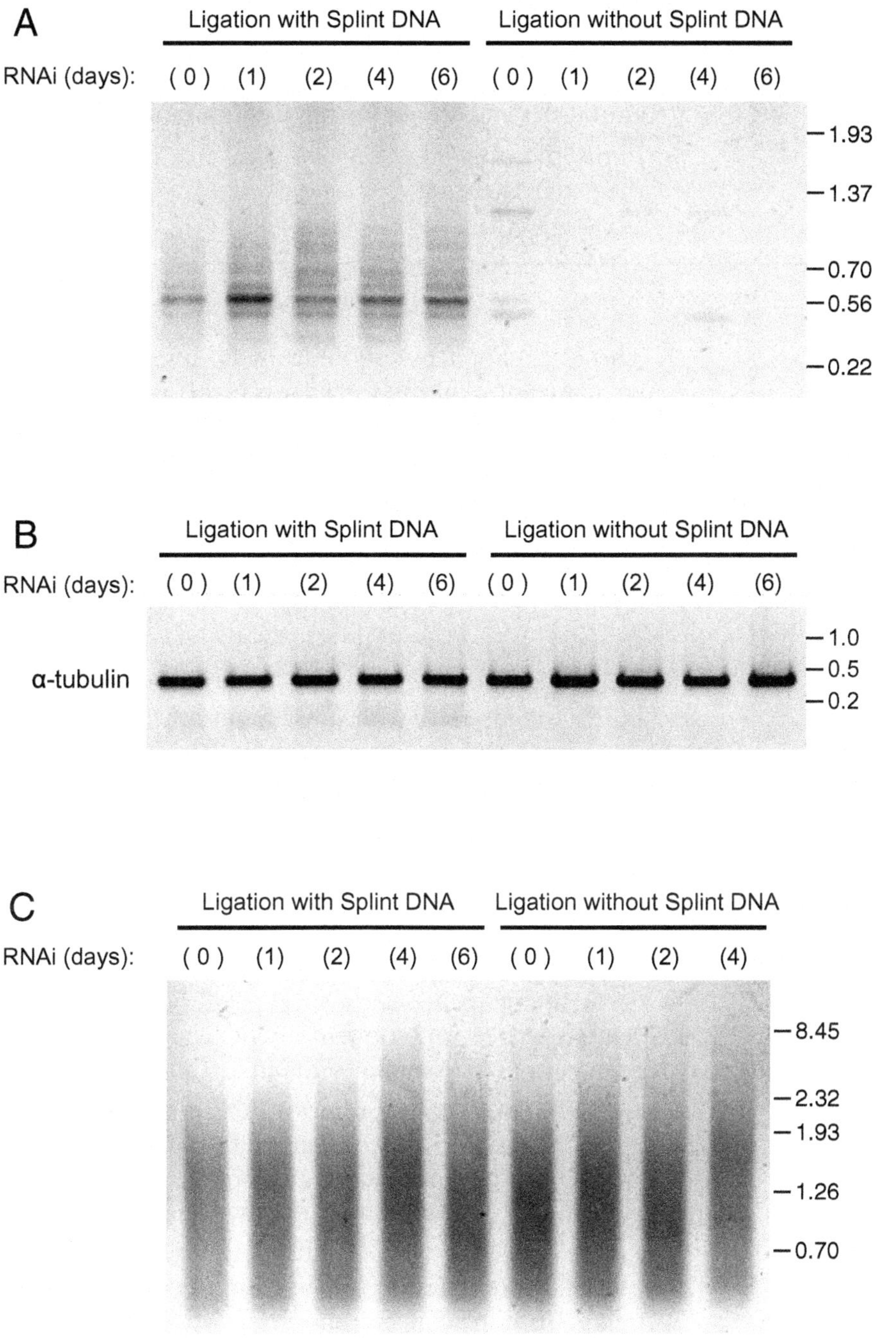

Fig. 2 Accumulation of 5′-PO_4 RNA in *Trypanosoma brucei* depleted for mRNA recapping enzyme (TbCe1). (**a**) *T. brucei* mRNAs isolated from TbCe1 RNAi uninduced (day 0) and induced cells (for the number of days indicated) were subjected to ligation-mediated RT-PCR. The RNA was converted to cDNA using oligo$(dT)_{20}$ as a primer, and splint ligation products were detected by PCR using primer specific to the 5′-anchor sequence and oligo$(dT)_{20}$. A series of PCR products, in the range of 0.4–1.2 kb, were detected. The amounts of splint ligation products, as well as their heterogeneity, were significantly higher in TbCe1-silenced than the unsilenced control. Control reactions in which the splint DNA was omitted were carried out in parallel. (**b**) The internal control reaction was performed using SL-specific forward primer and an α-tubulin-specific reverse primer. (**c**) Control reaction was performed using SL forward primer and oligo$(dT)_{20}$ primer. The positions of marker DNA (in kilobases) are indicated to the Right. Figure adapted from [6], Copyright (2015) National Academy of Sciences

4. Spread the cells on the Luria-Bertani plate with appropriate antibiotics and screen for the positive clones with insert.
5. Purify the plasmid DNA and sequence the inserted DNA by the Sanger method.

4 Notes

1. The 5′-anchor RNA should not be phosphorylated to avoid circularization during the ligation reaction.
2. The Agilent 2100 Bioanalyzer (Agilent Technologies) also provides a good estimate of RNA concentration and purity. The integrity and purity can be simultaneously analyzed in a single 5 ng sample.
3. The 5′-anchor RNA and splint DNA oligonucleotides should be mixed prior to the addition of the RNA sample to ensure that all the 5′-anchor RNA hybridize to the splint DNA.
4. Although RNA is most stable in TE buffer, the EDTA in the buffer will interfere with subsequent downstream RT-PCRs by chelating catalytic magnesium needed by the polymerase. Therefore, resuspend only in RNase-free water.
5. RNA in water stored at −20 °C should be used within a week. Storage at −70 °C extends life to about 1 month, but each freeze-thaw cycle reduces the quality of the subsequent cDNA and qRT-PCR results.
6. The oligo(dT) primer may be substituted with custom 3′-anchor sequence with 3′-dT_{20} tail in the reverse transcription reaction. Subsequent PCR can be performed with 5′-anchor and 3′-anchor sequence primers.
7. In this method, the amount of PCR product reaches a plateau and does not directly correlate with the amount of cDNA. To determine the amount of uncapped transcripts, quantitative real-time PCR should be performed.
8. GelRed™ or GelGreen™ (Biotium) can be used in place of ethidium bromide for visualization of nucleic acids in agarose gels. Use at 1× concentration into cooling agarose. These are low toxicity and highly stable compounds.
9. I recommend performing additional control PCR experiment using the 5′-anchor primer alone. If the PCR product is detected using forward primer alone, annealing temperature should be increased to eliminate nonspecific PCR amplification or alter the oligo(dT) primer used in the reverse transcription reaction.

Acknowledgment

I thank Bryan Mathis (University of Tsukuba) for scientific editing service. This work was supported by the National Science Foundation under Grant Number 1050984 and the JSPS Grants-in-Aid for Scientific Research KAKENHI 16H05180.

References

1. Topisirovic I, Svitkin YV, Sonenberg N, Shatkin AJ (2011) Cap and cap-binding proteins in the control of gene expression. Wiley Interdiscip Rev RNA 2:277–298. doi:10.1002/wrna.52
2. Liu H, Kiledjian M (2006) Decapping the message: a beginning or an end. Biochem Soc Trans 34:35–38. doi:10.1042/BST20060035
3. Grudzien-Nogalska E, Kiledjian M (2017) New insights into decapping enzymes and selective mRNA decay. Wiley Interdiscip Rev RNA 8, p. e1379. doi: 10.1002/wrna.1379.
4. Franks TM, Lykke-Andersen J (2008) The control of mRNA decapping and P-body formation. Mol Cell 32:605–615. doi:10.1016/j.molcel.2008.11.001
5. Schoenberg DR, Maquat LE (2009) Re-capping the message. Trends Biochem Sci 34:435–442. doi:10.1016/j.tibs.2009.05.003
6. Ignatochkina AV, Takagi Y, Liu Y et al (2015) The messenger RNA decapping and recapping pathway in Trypanosoma. Proc Natl Acad Sci U S A 112:6967–6972. doi:10.1073/pnas.1424909112
7. Wang LK, Nandakumar J, Schwer B, Shuman S (2007) The C-terminal domain of T4 RNA ligase 1 confers specificity for tRNA repair. RNA 13:1235–1244. doi:10.1261/rna.591807
8. Nandakumar J, Ho CK, Lima CD, Shuman S (2004) RNA substrate specificity and structure-guided mutational analysis of bacteriophage T4 RNA ligase 2. J Biol Chem 279:31337–31347. doi:10.1074/jbc.M402394200
9. Ho CK, Shuman S (2002) Bacteriophage T4 RNA ligase 2 (gp24.1) exemplifies a family of RNA ligases found in all phylogenetic domains. Proc Natl Acad Sci U S A 99:12709–12714. doi:10.1073/pnas.192184699
10. Takagi Y, Sindkar S, Ekonomidis D et al (2007) Trypanosoma brucei encodes a bifunctional capping enzyme essential for cap 4 formation on the spliced leader RNA. J Biol Chem 282:15995–16005. doi:10.1074/jbc.M701569200
11 Shenmin Yin, C Kiong Ho, Eric S Miller, Stewart Shuman (2004) Characterization of bacteriophage KVP40 and T4 RNA ligase 2. Virology 319(1):141–151. doi: 10.1016/j.virol.2003.10.037

Chapter 2

Preparation of Splicing Competent Nuclear Extract from Mammalian Cells and In Vitro Pre-mRNA Splicing Assay

Maliheh Movassat, Hossein Shenasa, and Klemens J. Hertel

Abstract

The ability to perform in vitro splicing assays has paved the way for in-depth studies of the mechanisms and machinery involved in the process of splicing. The in vitro splicing assay is a valuable experimental approach that combines the complexity of the spliceosome and regulatory systems with the flexibility of performing endless splicing and alternative splicing reactions. Through the use of crude nuclear extract and radiolabeled pre-mRNA, spliced mRNAs can be visualized using autoradiography for downstream analysis. This chapter describes the necessary steps to perform an in vitro splicing reaction, including the generation of the key components necessary for the splicing reaction; nuclear extract.

Key words Nuclear extract, HeLa cells, In vitro splicing, Splicing, Alternative splicing, mRNA processing, Pre-mRNA substrate, In vitro transcription, RNA extraction and purification

1 Introduction

RNA splicing is an essential co-transcriptional feature of gene expression in eukaryotes [1]. The spliceosome assembles in a stepwise manner on pre-mRNA transcripts as RNA polymerase elongates the nascent chain. RNA splicing entails the simultaneous excision and ligation of gene coding exons and the eventual degradation of introns [2]. Alternative splicing involves the selective excision and ligation of gene coding exons and is responsible for much of the genetic diversity seen in higher order organisms [2, 3].

In the field of RNA biology, cell-free or in vitro-based assays [4] have played a key role in providing mechanistic insights into the molecular workings of the cell. One such assay has been the in vitro splicing assay [5, 6], which allows for an experimenter controlled method for the investigation of splicing mechanisms, including

Maliheh Movassat and Hossein Shenasa contributed equally to this work.

Yongsheng Shi (ed.), *mRNA Processing: Methods and Protocols*, Methods in Molecular Biology, vol. 1648,
DOI 10.1007/978-1-4939-7204-3_2, © Springer Science+Business Media LLC 2017

spliceosomal assembly, splicing kinetics, and splicing regulatory processes. In vitro splicing reactions require two critical components: minigene constructs and nuclear extract. Minigene constructs are small fragments of DNA that, at minimum contain two exons, an intron and a phage promoter sequence upstream of the coding strand [7]. Minigene constructs can be transcribed from DNA to RNA using commercial phage polymerases in an in vitro transcription reaction. Nuclear extract preparations were originally developed to study RNA polymerase II transcription in the test tube [8]; however, these extracts were also shown to support intron excision [9]. Over the years many forms of splicing competent nuclear extract have been developed [10–15], all of which contain the essential components required for an efficient splicing reaction of pre-synthesized pre-mRNAs. The extract contains: spliceosomal components, splicing regulatory proteins, ATP, and other components of the nucleus [14, 16]. HeLa cells are the most common cells from which nuclear extract is prepared; however, nuclear extract from other cells can also be prepared in the same manner. In vitro splicing reactions generally use radiolabeled transcription of minigene constructs and subsequent incubation with nuclear extract [6, 16]. Proteins are digested using Proteinase K and the RNA is then separated from the proteins using phenol/chloroform extraction. After RNA precipitation, the spliced RNAs can be electrophoretically separated on a denaturing polyacrylamide gel and analyzed by autoradiography.

In vitro splicing reactions can be used to study many aspects of the splicing reaction such as splice site strength of the exon/intron junctions, influence of splicing regulatory elements, or the molecular interactions during splice site pairing. The ease with which the experimenter can manipulate the spliceosomal assembly pathway renders this assay an invaluable tool.

2 Materials and Reagents

All reagents and materials should be of high quality, RNase free, and molecular biology grade. In vitro splicing steps usually require work with radioactive isotopes; therefore, all necessary precautions must be taken. Carefully follow all local hazardous and radioactive waste disposal regulations when carrying out experiments using radiolabeled RNA.

2.1 Nuclear Extract Components and Reagents

2.1.1 Cells

Spinner cultured suspension HeLa-S3 cells (*see* **Note 1**).

2.1.2 Reagents

1. 1 M dithiothreitol (DTT).
2. 100 mM phenylmethanesulfonyl fluoride (PMSF) in isopropanol (*see* **Note 2**).

3. 1 M 4-(2-hydroxyethyl)piperazine-1-ethanesulfonic acid (HEPES) buffer, pH 7.9 at 4 °C with KOH. Store at 4 °C (*see* **Note 3**).
4. 1 M magnesium acetate ($Mg(OAc)_2$) (*see* **Note 4**).
5. 2.5 mM potassium acetate (KOAc) (*see* **Note 4**).
6. 0.5 mM EDTA.
7. 10× Phosphate-buffered saline (PBS), pH 7.4. Store at 4 °C.
8. Glycerol (at least 2 L needed for a 30 L culture of cells).
9. Autoclaved or double-deionized water.
10. Trypan blue.

2.1.3 Buffer Recipes

All reagents should be prepared with autoclaved or double-deionized water, followed by filter sterilization with a 0.22 μm filter or sterilized by autoclaving. All reagents should be stored at 4 °C and be cold prior to use. DTT and PMSF should be made fresh and only added to each corresponding buffer prior to use, not ahead of time.

1. Hypotonic buffer: 10 mM HEPES-KOH at pH 7.9, 1.5 mM $Mg(OAc)_2$, 10 mM KOAc, 0.5 mM DTT, 0.2 mM PMSF.
2. Low-salt buffer: 20 mM HEPES-KOH at pH 7.9, 1.5 mM $Mg(OAc)_2$, 20 mM KOAc, 0.2 mM EDTA, 25% glycerol, 0.5 mM DTT, 0.2 mM PMSF.
3. High-salt buffer: 20 mM HEPES-KOH at pH 7.9, 1.5 mM $Mg(OAc)_2$, 1.4 M KOAc, 0.2 mM EDTA, 25% glycerol, 0.5 mM DTT, 0.2 mM PMSF.
4. Dialysis buffer: 20 mM HEPES-KOH at pH 7.9, 100 mM KOAc, 0.2 mM EDTA, 20% glycerol, 0.5 mM DTT, 0.2 mM PMSF.

2.1.4 Equipment

All glassware (including dounce homogenizer) and bottles/tubes should be sterilized and autoclaved ahead of time. Make sure all glassware and plasticware that is used does not contain detergent residue.

1. Dialysis tubing: 10,000 molecular weight cutoff (MWCO).
2. Glass dounce homogenizer with a tight clearance pestle (*see* **Note 5**).
3. Centrifuge with swinging bucket rotor capable of speeds up to 3500 × *g*.
4. Centrifuge bottles: polypropylene, conical bottom with graduations, wide mouth with sealing caps.
5. Ultracentrifuge with fixed angle rotor capable of speeds up to 25,000 × *g*.
6. Centrifuge tubes: polycarbonate with polypropylene screw caps, 30 mL volume.

7. 25 mL serological glass pipettes.
8. 1 L glass bottles.
9. 200 mL glass beaker.
10. 4 L beaker/buckets for dialysis.
11. 1.5 mL tubes.
12. Column to aid in drip addition of high-salt buffer (optional).
13. Glass slide(s).
14. Phase-contrast microscope.
15. Magnetic stir bar.
16. Magnetic stir plate.
17. Ice.
18. Dry ice.
19. −80 °C Freezer.

2.2 In Vitro Splicing Reaction

2.2.1 Splicing Reaction Components

1. Radiolabeled pre-mRNA generated from an in vitro transcription reaction (*see* **Note 6**).
2. Splicing competent nuclear extract (*see* Subheading 3.1).
3. 25 mM adenosine triphosphate (ATP).
4. 0.5 M creatine phosphate (CP).
5. 80 mM $Mg(OAc)_2$ (*see* **Note 4**).
6. RNase inhibitor, 40 units/μL.
7. 100 mM DTT.
8. 13% polyvinyl alcohol (PVA) (optional).
9. 1 M KOAc (*see* **Note 4**).
10. 0.5 M HEPES-KOH, pH 7.9.
11. Wet ice and dry ice.
12. Water bath.

2.2.2 Splicing gel Components

1. Upright vertical gel electrophoresis system.
2. Electrophoresis power supply with temperature probe.
3. 8″ × 8″ glass plates or equivalent.
4. 0.4 mm spacer set.
5. 0.4 mm gel comb.
6. Aluminum plate: 8″ × 8″ or longer and precooled.
7. 1¼″ binder clips (at least four).
8. 10× Tris-Borate-EDTA (TBE): 1 M Tris Base, 1 M boric acid, 20 mM EDTA.
9. 7 M Urea prepared in 1× TBE.
10. 20% (19:1) acrylamide:bis-acrylamide solution: 210.2 g solid urea, 50 mL 10× TBE, 250 mL 40% acrylamide:bis-acrylamide

solution, adjust to 500 mL with sterile water to yield a final concentration of 20% acrylamide/7 M urea/1× TBE.

11. *N,N,N′,N′*-Tetramethylethylenediamine (TEMED).
12. 10% ammonium persulfate (APS).
13. Stop dye: 98% formamide, 0.1% bromophenol blue, 0.1% xylene cyanol, 10 mM EDTA at pH 8.
14. Silicone-based coating solution.
15. 70% ethanol.
16. Putty knife/gel spatula.
17. Filter paper cut in 8″ × 8″ squares.
18. Plastic wrap.
19. Gel dryer.
20. PhosphorImager System. Film may also be used.
21. Lint-free tissue.
22. 30 mL syringe and needles.

2.2.3 RNA Purification

1. Proteinase K at 10 mg/mL.
2. 2× Proteinase K buffer: 20 mM Tris Base, 2% SDS, 200 mM NaCl, 2 mM EDTA, pH 7.5.
3. 100% ethanol.
4. Glycogen.
5. Phenol, chloroform, isoamyl alcohol solution (25:24:1 pH 8.0).

3 Methods

3.1 Nuclear Extract Preparation

To prevent the denaturation of RNA and proteins, all the extraction steps should be carried out on ice (or in a coldroom). All reagents and buffers should be equilibrated to 4 °C and centrifuge rotors should be precooled to 4 °C. Once initiated, this protocol should be carried all the way to completion. *See* Fig. 1 for a quick reference guide to nuclear extract preparation.

3.1.1 Isolation of Nuclei

1. Transfer HeLa cells to a conical centrifuge bottle to pellet the cells. Centrifuge at 1000 × *g* for 5 min at 4 °C. Decant the supernatant carefully so as not to disturb the pellet.
2. Wash the cells by resuspending the pellet with ice-cold 1× PBS. Add PBS at 5× the volume of cells (*see* **Note 7**).
3. Pellet cells post-wash by centrifugation at 1850 × *g* for 10 min at 4 °C. Remove the supernatant carefully. From these pelleted cells, determine the packed cell volume (PCV), using the graduations on the centrifuge bottle.
4. Wash the cells GENTLY by resuspending the pellet with 5× PCV hypotonic buffer. Immediately centrifuge cells to pellet

Nuclear Extract Quick Reference Guide

Date:____________
Start Time:__________
End Time:__________

Cell Type____________ Cell Count____________
Culture Volume____________ Total # of Cells:____________

Isolation of Nuclei

- ☐ Collect and centrifuge culture cells: 1,000x*g*, 5 min, 4°C.
- ☐ Wash cells: 1X PBS – collect in conical tube.
- ☐ Centrifuge: 1,850x*g*, 10 min, 4°C.
- ☐ Determine packed cell volume (PCV):____________
- ☐ Add DTT and PMSF to *hypotonic buffer.*
- ☐ Wash cells: 5x PCV, *hypotonic buffer.* Volume used:____________
- ☐ Centrifuge: 1,850x*g*, 10 min, 4°C.
- ☐ Resuspend to 3x original PCV, *hypotonic buffer.* Volume used:____________
- ☐ Incubate on ice, 10 min.
- ☐ Check for cell lysis. % Lysed:____________
- ☐ Homogenize cells (dounce): 10-20 strokes/plunges.
- ☐ Check for cell lysis. % Lysed:____________

Extraction of Nuclei

- ☐ Pellet nuclei, centrifuge: 3,000x*g*, 15 min, 4°C.
- ☐ Determine packed nuclear volume (PNV):____________
- ☐ Optional: Save supernatant for S-100 preparation.
- ☐ Add DTT and PMSF to *low-salt buffer.*
- ☐ Resuspend pellet to 0.5x PNV, *low-salt buffer.* Volume used:____________
- ☐ Transfer nuclei to glass beaker with stir bar.
- ☐ Add DTT and PMSF to *high-salt buffer.*
- ☐ Add 0.5x PNV, *high-salt buffer*, drop-wise. Volume used:____________
- ☐ Lyse nuclei: stir on ice, 30 minutes.
- ☐ Transfer to tubes and centrifuge: 25,000x*g*, 30min, 4°C.
- ☐ SAVE Supernatant = Nuclear Extract!

Dialysis and Storage of Extract

- ☐ Rinse dialysis tubing.
- ☐ Add DTT and PMSF to *dialysis buffer.*
- ☐ Dialyze extract (3x, 1.5 hours, 4°C): 50x supernatant volume, *dialysis buffer.*
  Total volume needed per change:____________
  1) Start time:____________Stop time:____________
  2) Start time:____________Stop time:____________
  3) Start time:____________Stop time:____________
- ☐ Centrifuge: 25,000x*g*, 30 min, 4°C.
- ☐ Volume of nuclear extract (supernatant):____________# of Aliquots:____________
- ☐ Store -80°C.

Fig. 1 Nuclear extract preparation guide for quick referencing and data recording

at 1850 × *g* for 10 min at 4 °C and decant the supernatant (*see* **Note 8**).

5. Resuspend the packed cells in hypotonic buffer to a final volume of 3× the original PCV (the volume of the cells and the hypotonic buffer combined should be 3× PCV). Incubate cells on ice and allow the cells to swell for 10 min (*see* **Note 9**).
   (a) Check for cell lysis of pre-dounced cells. Lysis can be determined by visualizing stained cells under the microscope by the addition of trypan blue. For a dense concentration of cells, dilute with 1× PBS (*see* **Note 10**).
6. Transfer the cells to a dounce homogenizer to aid in cell lysis. Homogenize the cells with 10–20 plunges/strokes using an up and down motion (*see* **Note 11**).
   (a) Monitor cell lysis by checking post-dounced cells as previously described in **step 5a** (*see* **Note 12**).

3.1.2 Extraction of Nuclei

1. Transfer dounced cells to clean centrifuge bottles and spin to pellet by centrifugation at 3300 × *g* for 15 min at 4 °C. Determine and record the packed nuclear volume (PNV), using the graduations on the centrifuge bottle. At this point the supernatant can be saved for cytoplasmic S-100 extract preparation (*see* **Note 13**).
2. Resuspend the pellet (which now contains nuclei) by adding 0.5× PNV low-salt buffer. Transfer the resuspension to a glass beaker with a magnetic stir bar and gently stir on ice (*see* **Note 14**).
3. Gently release the soluble proteins from the nuclei by adding 0.5× PNV high-salt buffer in a dropwise fashion. Continue to stir on ice for 30 min to complete the extraction of the nuclei (*see* **Note 15**).
4. Transfer nuclei to centrifuge tubes to pellet by centrifugation at 25,000 × *g* for 30 min. Save the supernatant, as this is the nuclear extract.

3.1.3 Dialysis and Storage of Extract

1. Prepare the dialysis tubing by rinsing in distilled water while samples are in the centrifuge (*see* **Note 16**).
2. Desalt the nuclear extract by dialyzing the supernatant in the dialysis tubing dialysis buffer at 50× supernatant volume. Dialyze for 1.5 h at 4 °C, while stirring (*see* **Note 17**).
3. Change the dialysis buffer two additional times and dialyze for 1.5 h at 4 °C each time.
4. Transfer the nuclear extract to centrifuge tubes and remove the precipitate by centrifugation at 25,000 × *g* for 30 min at 4 °C. Save the supernatant.

5. Aliquot the supernatant into 1 mL fractions and freeze on dry ice immediately (*see* **Note 18**).
6. Store the nuclear extract aliquots at −80 °C (*see* **Note 19**).
7. Validate the activity of the nuclear extract by performing an in vitro splicing reaction.

3.2 In Vitro Splicing Assay

3.2.1 Splicing Reaction

1. Determine the volume for a master mix. The master mix volume can be determined with the following equation:
   Reaction volume × (# of reactions + 1) = master mix volume
   (a) For example, 25 μL × (4 reactions + 1) = 125 μL. The extra (+1) volume is to account for pipetting errors. Typical splicing reaction volumes range between 10 and 25 μL.
2. Thaw nuclear extract on ice.
3. Allow the following reagents to thaw at room temperature and then immediately place them on ice: ATP, CP, $Mg(OAc)_2$, DTT, HEPES, and KOAc. Thaw radiolabeled RNA and nuclear extract on ice. Keep RNase inhibitor on ice as well (*see* **Note 4**).
4. Combine the following reagents to obtain a final concentration of: 1 mM ATP, 20 mM CP, 3.2 mM $Mg(OAc)_2$, 10 units RNase inhibitor, 1 mM DTT, 3% PVA (optional), 12 mM HEPES, 72.5 mM KOAc, and 10–50% nuclear extract (percent of nuclear extract to use should be optimized for each extract and substrate that will be used). Add sterile water to bring up the master mix volume if necessary (*see* **Notes 20–22**).
5. For each experimental condition, add the following: appropriate volume of master mix, 0.01–0.1 nM RNA (~1000 cpm), experimental variant (such as protein), and/or sterile water to bring up the volume. Add nuclear extract last to initiate the reaction. Gently pipette up and down to mix the reaction. Keep all reaction tubes on ice. Once nuclear extract has been added, quickly take a time 0 ($T = 0$) aliquot from each reaction tube and immediately freeze on dry ice (*see* **Note 23**).
6. Incubate all reactions (except the time 0 control) at 30 °C for 90 min in a water bath (*see* **Note 24**).
7. While the splicing reaction is incubating, prepare the 6% polyacrylamide gel.

3.2.2 Splicing Gel Preparation

1. Prepare a 6% polyacrylamide solution in a 50 mL conical tube. For the gel size specified above, 25 mL is sufficient. Dilute 20% acrylamide/7 M urea/1× TBE with 7 M urea/1× TBE to obtain a 6% acrylamide solution.
2. Carefully clean the glass plates by running them under deionized water (*see* **Note 25**).

3. Place both glass plates on a flat surface with the inside of each plate facing upward. Wipe away any residual water with lint-free paper towels.
4. Spray the glass plates with 70% ethanol and wipe them dry with lint-free paper towels (*see* **Note 26**).
5. Assemble the gel cassette (*see* **Note 27**) and add binder clips to the edges and sides of the glass plates to hold the plates together, taking care to not move the spacers (*see* **Note 28**).
6. Once the gel cassette is ready, add a 1:1000 volume of TEMED and a 1:100 volume of 10% APS to the 6% acrylamide solution (*see* **Note 29**).
7. Aspirate the polyacrylamide solution with a 30 mL syringe that does not have a needle. Hold the gel at a 45° angle and place the syringe tip such that it makes firm and direct contact with the non-siliconized plate. Apply constant pressure to dispense the acrylamide solution. Once the cassette is full, place it on a flat elevated surface such as a test tube rack (*see* **Note 30**).
8. Insert the gel comb (*see* **Note 31**).
9. Leave the cassette flat on a bench top. Flush out any remaining un-polymerized gel from the syringe back into the conical tube. This can be used as a marker to confirm that the gel has polymerized. Let the gel polymerize for ~30 min.
10. Remove the gel comb and bottom spacer. Clamp the gel cassette to the upright gel electrophoresis apparatus using 1¼″ binder clips (*see* **Note 32**).
11. Tilt the electrophoresis apparatus to one side, making a 45° angle between the bench top and the bottom of the electrophoresis apparatus, slowly pour 1× TBE buffer down the raised end while incrementally lowering the apparatus. This will minimize the formation of bubbles between the glass plates (*see* **Note 33**).
12. Make sure the bottom of the gel is submerged in buffer. Fill the top compartment of the electrophoresis apparatus with 1× TBE, until the wells in the gel are filled with buffer.
13. Prerun the gel for 15 min at 30 W. Attach a temperature probe and set the temperature limit to 45 °C to ensure the glass plates do not break.

3.2.3 Proteinase K Digest

1. Once the splicing reaction has reached completion, immediately place the tubes on dry ice.
2. Determine the final volume for the Proteinase K master mix using the following equation:

   Reaction volume × (# of reactions + 1) = master mix volume
3. Mix reagents to yield a final master mix volume: 1× Proteinase K buffer at final desired volume, 0.25 mg/mL glycogen (*see* **Note 34**) and 0.25 mg/mL Proteinase K. Use sterile water to

adjust the volume of the master mix (if necessary) to a final volume of 180 μL per reaction.

4. Add 175 μL of Proteinase K master mix to each tube and incubate at 37 °C for 10–15 min (*see* **Note 35**).

3.2.4 RNA Purification

1. Once the Proteinase K digest is complete, add 200 μL of phenol:chloroform to each reaction tube.
2. Vortex the tubes on high speed for 30s and centrifuge at 16,500 × *g* for 5 min.
3. Carefully remove the aqueous phase (top layer), taking care not to remove any of the organic phase.
4. Pipette the aqueous phase of each tube into a fresh tube and add 3 volumes of 100% ice-cold ethanol. Incubate the tubes at −20 °C for 15 min.
5. Centrifuge the tubes at 16,500 × *g* for 10 min at room temperature (*see* **Note 36**).
6. Remove the ethanol supernatant, taking care not to disturb the pellet. Allow the pellet to air dry for no more than a few minutes. Add a small volume of stop dye (10 μL or less) and pipette up and down to mix. At this point the RNA is ready to be loaded onto the gel.

3.2.5 Electrophoresis and Visualization

1. Load samples onto the gel and run at 30 W (100 V), 45 °C, for 60–90 min or until the bromophenol blue dye reaches the bottom of the gel. The length of time samples that are run on the gel should be optimized based on the size of expected products (*see* **Notes 37** and **38**).
2. Remove the buffer from the top and bottom compartments of the upright electrophoresis apparatus as well as the temperature probe. Carefully detach the 1¼″ binder clips.
3. Cut filter paper into 8″ × 8″ pieces.
4. Pull the vertical spacers out of the gel cassette and use the putty knife to wedge open the glass plates. Remove the siliconized (notched) plate.
5. Place a piece of 8″ × 8″ filter paper on the gel and gently press down. Invert the gel such that the filter paper is on the bench and the glass plate is on top.
6. Use the putty knife to remove the remaining glass plate by slowly lifting one corner. The gel should adhere to the filter paper.
7. Place a piece of plastic wrap on the top of the gel. Take care to avoid creases in the wrap.
8. Dry the gel for 20 min at 80 °C using a gel dryer. Make sure the suction pump is turned on for the duration of drying.
9. Expose a phosphorscreen to the gel for 1 h-overnight (*see* **Notes 39** and **40**).

10. Use a PhosphorImager to obtain an image of the spliced products. This image can be used to quantify parameters such as percentage of splicing or splice site preference in a minigene that contains competitive splice sites.
11. Quantify results using a gel analysis software (*see* **Note 41**).

4 Notes

1. This nuclear extract protocol starts with a large volume (~30 L) of purchased spinner cultured HeLa cells. HeLa cells can also be cultured in the lab.
2. PMSF is dissolved in anhydrous isopropanol and should be prepared fresh and added to corresponding buffers just prior to use (similar to addition of DTT). Store the freshly prepared PMSF solution on ice or at 4 °C during duration of the nuclear extract prep.
3. HEPES solution should be brought to pH 7.9 while at 4 °C or on ice.
4. $MgCl_2$ and KCl can be used as well; however, a previous study has shown that the use of acetate as a counter ion enhances the splicing reaction [17].
5. Use a pestle that has a tight fit in the mortar (approximately 0.025–0.076 mm): Kontes brand homogenizers have a tight clearence in their type-B pestle, alternatively, Wheaton dounce homogenizers have a tighter clearance in their type-A pestle.
6. Pre-mRNAs are generally transcribed with commercially available phage polymerases in the presence of UTP that contains phosphorous-32 at the alpha position (^{32}P α-UTP). The radioactive nucleotide is generally in the 0.3–3 nM range and usually leads to 100,000 cpm/μL incorporation. Other nucleotide triphosphates that have a radioactive α-phosphate can be used instead of ^{32}P α-UTP if needed.
7. It is suggested to use 25 mL serological glass pipettes to wash the cells. Make sure to mix gently and not expel volume from the pipette completely while mixing, so as to prevent the creation of a vacuum.
8. This step should be performed quickly. The hypotonic buffer swells the cells and could potentially cause the cells to leak or burst, leading to loss of protein into the supernatant or cell death. Monitoring cell lysis with trypan blue allows for visualization of lysed cells, which take up the dye, as compared to intact cells, which do not.
9. The previous wash step with hypotonic buffer may have initiated swelling of the cells. Therefore, the PCV may have increased. When determining how much hypotonic buffer to

add in this step, refer only to the initial PCV that was recorded. For example, the PCV determined in **step 4** is 15 mL, yet after **step 5** it has increased to 25 mL. In **step 6** add hypotonic buffer such that the final volume of cells and buffer is 45 mL (3 × 15 mL). Add the hypotonic solution gently while mixing with a serological pipette.

10. Check for the lysis of pre-dounced cells by adding trypan blue, 1:2 dilution. At this point, you want a minimum of 50% of the cells alive. Greater than 80% cell survival is ideal. Use a plastic pipette with a cut tip to check for lysis to prevent shearing of the cells.
11. Perform the dounce homogenization step with gentle strokes maintaining a constant plunging motion. Do not remove the pestle from the dounce until douncing is complete. This will ensure efficient cell breakage.
12. Visualize cell lysis with a 1:2 dilution of cells to trypan blue. 80–90% cell lysis should be expected. A good sampling of the lysed cells can be found from further down the dounce tube, and not from the top.
13. S-100 cytoplasmic fraction extraction can be performed at this step. For further instructions, *see* ref. 11.
14. The lysate can be homogenized again by douncing if the cell solution is chunky. At this point, if there are multiple tubes, combine the nuclei into one beaker.
15. The dropwise addition of the high-salt buffer is vitally important because rapidly increasing the salt concentration may lead to nuclear lysis and precipitation of nuclear components. High-salt buffer permeabilizes the nuclear membrane to allow for release of necessary components. Nuclei can be further homogenized with 5 dounce strokes to prevent clumping when nuclei are in large volumes.
16. Alternatively, dialysis tubing can be rinsed in dialysis buffer.
17. Avoid the presence of bubbles in the dialysis tubing by gently squeezing the bubbles out of the tubing with your fingers. Remember to clamp one end of the tubing prior to addition of sample.
18. 30 L of a HeLa cell culture with a 4–6 × 10^5 cells/mL density should yield about 45 mL of nuclear extract.
19. Freeze/thaw cycles of the nuclear extract aliquots should be limited to avoid compromising extract activity. The non-thawed extracts can be stored up to 2 years at −80 °C without loss of activity. However, the half-life at 4 °C is only 12 h [18].
20. The optimum potassium concentration for an in vitro splicing reaction is around 30 mM; however, splicing can be observed between 2 mM up to 100 mM. Splicing efficiency is reduced

drastically at the high and low extremes. Splicing efficiency may be increased by optimizing the potassium ion concentration [9].

21. When adding KOAc and HEPES it is important to account for the potassium ion and HEPES concentration already present in the nuclear extract. For example, if the nuclear extract already contains 100 mM KOAc and 20 mM HEPES it is possible to calculate how much additional KOAc and HEPES to add to the reaction with the following calculation:

$$\begin{gathered} \textit{For a } 25\,\mu L \textit{ reaction containing } 30\%\ NE: \\ (7.5\ \mu L\, NE)(100\ mM\ KOAc) = (25\ \mu L)(\mathrm{X}) \\ \mathrm{X} = 30\, mM\ KOAc. \end{gathered}$$

The addition of nuclear extract to the reaction yields an initial 30 mM KOAc concentration. This amount must be subtracted from the desired final concentration

$$\begin{gathered} 72.5\ mM\ KOAc - 30\ mM\ KOAc = 42.5\ mM\ KOAc \\ (1000\ mM\ KOAc)(\mathrm{X}) = (25\,\mu L)(42.5\ mM\ KOAc) \\ \mathrm{X} = 1.06\,\mu L\, KOAc. \end{gathered}$$

PVA is a concentration-enhancing polymer that may increase reaction efficiency. It may increase splicing, but is not essential to the reaction [19].

22. Taking a $T = 0$ aliquot and freezing it on dry ice stalls the reaction from proceeding. The $T = 0$ aliquot can be used as a zero time point to mark the initiation of the reaction.

23. Reaction times can be varied and optimized depending on the experimental conditions.

24. If there is residual debris stuck to the plates, water will be forced to flow around the debris. Hold the gel at a 45° angle and run water over it. Move glass plates slowly and scan for debris. Gently wipe debris away if any is present.

25. Cleaning glass plates with 100% ethanol will lead to faster removal of the silicone based coating; therefore, it is best to use 70% ethanol.

26. Place two spacers on the vertical edges of the non-siliconized glass plate. Make sure the spacers are aligned. Place the third spacer horizontally at the bottom of the glass plate. Make sure spacers are flush with each other and place the second (notched) glass plate onto the first. Make sure the siliconized side is facing toward the inside of the gel cassette.

27. If the spacers move during the attachment of 1¼″ binder clips, gaps can appear between them, which will lead to leakage from the gel cassette.

28. Once APS and TEMED have been added it is important to work quickly because the acrylamide will start to solidify. Invert the conical tube three times to ensure mixing.
29. It is critical to avoid bubbles. If a bubble appears inside the gel, tilt the gel cassette to one side and then bring it back to its original position with a rapid and continuous movement.
30. For the best results, press the gel comb flat against the non-siliconized glass plate. Insert the comb by placing thumbs on both sides of the comb and gently pushing down. Do not push the comb down past the top of the wells.
31. To remove the comb, hold the gel at a 45° angle (alternatively, place the gel flat on raised surface such as test tube rack) and place both thumbs at the edges of the comb. Gently push upward, taking care to distribute force evenly.
32. Ensure there are no bubbles on the bottom of the gel cassette. The presence of bubbles will cause uneven current distribution and may cause the gel box to shut off. Take a syringe with a bent needle and draw up buffer. Slide the needle along the bottom of the gel cassette in the opening between the two glass plates and simultaneously dispense buffer with minimal force to move the bubbles out.
33. Glycogen is polysaccharide carrier that aids in the precipitation and visualization of nucleic acids in ethanol precipitation protocols.
34. For 25 μL reactions, add 175 μL of Proteinase K master mix; however, this amount should be scaled up or down depending on the volume of the splicing reaction used.
35. Place tubes in a uniform fashion so that the pellet will appear on the same side for each tube. It is important to minimize loss of the RNA during the ethanol precipitation. If tubes are placed in a uniform manner such that the pellet appears in the same spot, the experimenter can assume the pellet is there even if it is faint or not visible.
36. Wash out the wells of the gel prior to loading samples to remove any residual urea that has settled at the bottom of the wells. Using a syringe with a straight needle attached aspirate some buffer from the top compartment of electrophoresis apparatus and dispense buffer into the wells with mild force to displace the urea.
37. Use 1¼″ binder clips to clamp a precooled aluminum plate to the front of the gel cassette. This will help dissipate heat and prevent the glass plates from breaking.
38. Phosphor storage technology makes use of phosphorscreens that are composed of $BaFBr:Eu^{2+}$ crystals immobilized in an organic matrix. Phosphorscreens can be used for autoradiography and have many advantages over the traditional method of

exposing x-ray film to radioactive gels. For example, phosphorscreens have between 10 and 250 times higher sensitivity and a linear dynamic range that spans five orders of magnitude. High-energy radiation emitted from radioactive atoms, such as ^{32}P, oxidizes Eu^{2+} to Eu^{3+} and leads to an electron being trapped in the BaFBr complex. The reduced $BaFBr^{-}$ has a unique absorbance in the 600 nm range. Exposure of the phosphorscreen to a 633 nm wavelength scanning laser results in the oxidation of the $BaFBr^{-}$ complex and reduction of Eu^{3+} back to Eu^{2+}. The reduction of Eu^{3+} to Eu^{2+} leads to the emission of a photon with a wavelength of 390 nm. These photons can be detected with a photomultiplier instrument as the laser scans the phosphorscreen [20].

39. A 1 h exposure to the phosphorscreen is sufficient for pre-mRNA that has a specific activity of 800 cpm/μL or higher. Longer exposure times may be needed to visualize less abundant splicing intermediates; an optimum exposure time can be found through trial and error.

40. There are many forms of analysis for autoradiograms of in vitro splicing reactions. One of the simplest forms of analysis is to calculate the percent of spliced product (Fig. 2). Calculate the sum of the volume intensity for the spliced band and the unspliced band taking into account background and time 0 (total signal). Divide the intensity of the band corresponding to the spliced product by the total signal intensity and then multiply by 100 to obtain a percentage.

$$\%\ Spliced = \frac{Signal\ from\ final\ spliced\ product}{Total\ signal\ in\ lane}$$

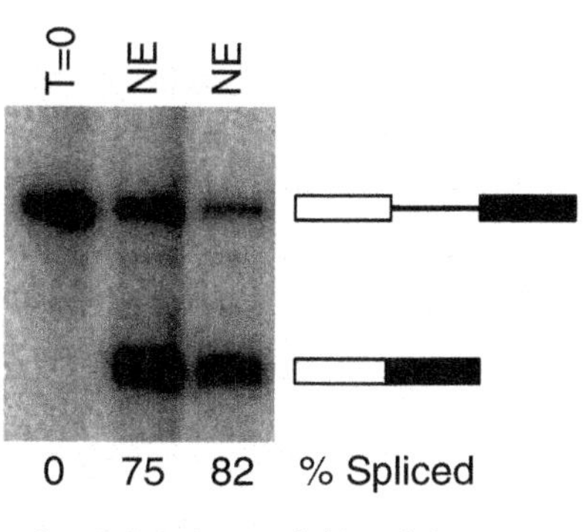

Fig. 2 Autoradiogram of radiolabeled β-globin minigene pre-mRNA splicing reaction (from *left* to *right*) at time 0 (*T* = 0) (*lane 1*) or incubated for 90 min in 30% nuclear extract (NE) from two different extract preparations (*lane 2* and *3*). Reaction products were separated on a 6% polyacrylamide gel. The splicing efficiency (% spliced) was determined using software from Bio-Rad Quantity One® (*see* **Note 41**)

Acknowledgments

Research in the Hertel laboratory is supported by NIH (GM062287 and GM110244). Maliheh Movassat and Hossein Shenasa contributed equally to this work.

References

1. Merkhofer EC, Hu P, Johnson TL (2014) Introduction to cotranscriptional RNA splicing. Methods Mol Biol 1126:83–96
2. Moore MJ, Query CC, Sharp PA (1993) 13 Splicing of precursors to mRNA by the Spliceosome. Cold Spring Harb Monogr Arch 24:303–357
3. Hertel KJ (2008) Combinatorial control of exon recognition. J Biol Chem 283: 1211–1215
4. Roca X, Karginov FV (2012) RNA biology in a test tube – an overview of in vitro systems/ assays. Wiley Interdiscip Rev RNA 3:509–527
5. Hicks MJ, Lam BJ, Hertel KJ (2005) Analyzing mechanisms of alternative pre-mRNA splicing using in vitro splicing assays. Methods 37:306–313
6. Movassat M, Mueller WF, Hertel KJ (2014) In vitro assay of pre-mRNA splicing in mammalian nuclear extract. Methods Mol Biol 1126:151–160
7. Cooper TA (2005) Use of minigene systems to dissect alternative splicing elements. Methods 37:331–340
8. Dignam JD, Lebovitz RM, Roeder RG (1983) Accurate transcription initiation by RNA polymerase II in a soluble extract from isolated mammalian nuclei. Nucleic Acids Res 11:1475–1489
9. Krainer AR, Maniatis T, Ruskin B, Green MR (1984) Normal and mutant human β-globin pre-mRNAs are faithfully and efficiently spliced in vitro. Cell 36:993–1005
10. Pugh BF (1995) Preparation of HeLa nuclear extracts. Methods Mol Biol 37:349–357
11. Mayeda A, Krainer AR (1999) Preparation of HeLa cell nuclear and cytosolic S100 extracts for in vitro splicing. Methods Mol Biol 118:309–314
12. Abmayr SM, Yao T, Parmely T, Workman JL (2006) Preparation of nuclear and cytoplasmic extracts from mammalian cells. Curr Protoc Mol Biol Chapter 12:Unit 12.1
13. Kataoka N, Dreyfuss G (2008) Preparation of efficient splicing extracts from whole cells, nuclei, and cytoplasmic fractions. Methods Mol Biol 488:357–365
14. Webb C-HT, Hertel KJ (2014) Preparation of splicing competent nuclear extracts. Methods Mol Biol 1126:117–121
15. Nilsen TW Preparation of nuclear extracts from HeLa cells. Cold Spring Harb Protoc 2013, 2013:579–583
16. Hernandez N, Keller W (1983) Splicing of in vitro synthesized messenger RNA precursors in HeLa cell extracts. Cell 35:89–99
17. Reichert V, Moore MJ (2000) Better conditions for mammalian in vitro splicing provided by acetate and glutamate as potassium counterions. Nucleic Acids Res 28:416–423
18. Carey MF, Peterson CL, Smale ST (2009) Dignam and Roeder nuclear extract preparation. Cold Spring Harb Protoc 4:1–4
19. Mayeda A, Krainer AR (1999) Mammalian in vitro splicing assays. Methods Mol Biol 118:315–321
20. Johnston RF, Pickett SC, Barker DL (1990) Autoradiography using storage phosphor technology. Electrophoresis 11:355–360

Chapter 3

Single-Molecule Analysis of Pre-mRNA Splicing with Colocalization Single-Molecule Spectroscopy (CoSMoS)

Joerg E. Braun and Victor Serebrov

Abstract

Recent development of single-molecule techniques to study pre-mRNA splicing has provided insights into the dynamic nature of the spliceosome. Colocalization single-molecule spectroscopy (CoSMoS) allows following spliceosome assembly in real time at single-molecule resolution in the full complexity of cellular extracts. A detailed protocol of CoSMoS has been published previously (Anderson and Hoskins, Methods Mol Biol 1126:217–241, 2014). Here, we provide an update on the technical advances since the first CoSMoS studies including slide surface treatment, data processing, and representation. We describe various labeling strategies to generate RNA reporters with multiple dyes (or other moieties) at specific locations.

Key words CoSMoS, Colocalization single-molecule spectroscopy, Pre-mRNA splicing, Spliceosome, TIRF microscopy, Klenow fragment, RNA ligation, Rastergram, Single-molecule spectroscopy, Single-molecule fluorescence microscopy

1 Introduction

Pre-mRNA splicing is a highly complex process and even 40 years after its discovery [1, 2] key mechanistic aspects remain elusive. Ensemble methods are powerful to characterize stable intermediates of the splicing reaction and the properties of the spliceosomal components, but are not ideally suited to study such a multi-step process with alternative pathways. Single-molecule techniques however can overcome many limitations of classical ensemble methods and provide insights into reaction intermediates, alternative pathways and define their kinetics [3]. Remarkable progress has been made in the past decade to apply single-molecule techniques to biological systems [4].

In the last decade single-molecule approaches have been developed and applied to study the dynamics of splicing and the

Yongsheng Shi (ed.), *mRNA Processing: Methods and Protocols*, Methods in Molecular Biology, vol. 1648,
DOI 10.1007/978-1-4939-7204-3_3, © Springer Science+Business Media LLC 2017

spliceosome [5–12]. All experiments involve immobilization of pre-mRNA reporter on the slide surface and labeling molecules of interest with fluorescent dyes. The fluorescence intensity at the reporter locations on the slide surface is observed with total internal reflection fluorescence (TIRF) microscopy. To obtain information on the comings and goings of molecules of interest the fluorescence intensity is monitored over time, which has been termed colocalization single-molecule spectroscopy (CoSMoS). To obtain information on the distance of two dyes to each other, their Förster resonance energy transfer (FRET) signal is monitored over time. The above-mentioned single-molecule studies of splicing employ either CoSMoS or FRET or combinations of these two strategies.

In this chapter, we focus on the CoSMoS technique, which allows observing comings and goings of molecules in real time [7]. A detailed protocol of CoSMoS was published recently [13]. Since the first CoSMoS studies of splicing in the complexity of cellular extracts several technical advances have been made, especially regarding preparation of reagents and data processing. Therefore, we provide here an update on how to prepare pre-mRNA reporters and outline updated data processing and representation.

We discuss two strategies that allow incorporating multiple dyes at specific locations in a pre-mRNA reporter. Using Klenow extension, multiple dyes or other moieties can be added at the 3′ end of RNAs as a straightforward way to generate bright pre-mRNA reporters. Another strategy is based on incorporation of modified nucleotides during the in vitro transcription reaction to allow subsequent dye-labeling. At the end of this chapter, we share notes on best practices for successful CoSMoS experiments. At the end of this chapter, we have notes not only regarding pre-mRNA reporter preparation, but all steps of CoSMoS experiments. The notes cover points whose importance has emerged over the last years, including treatment of the slide surface, a quantitative internal control for nonspecific binding to the slide surface using control locations, and rastergrams as a powerful visual tool to represent CoSMoS data (*see* **Notes 9–11**).

2 Materials

2.1 3′ End Addition of Multiple Dyes to RNA Using Klenow

1. DNA template for Klenow extension.
2. Klenow fragment of DNA polymerase I (3′-5′ exo-).
3. 10× concentrated annealing buffer: 100 mM HEPES (pH 7.5), 200 mM NaCl, 1 mM EDTA.
4. NEBuffer 2 (New England Biolabs).
5. dNTPs.

6. Dye- or biotin-labeled dUTP or dCTP.
7. RNase-free H_2O.
8. DNA template trap.
9. 0.5 M EDTA.
10. 5 M ammonium acetate.
11. 100% Ethanol.

2.2 Multi-dye Body-Labeled RNA Oligomer Preparation

1. Linear DNA template for in vitro transcription (*see* **Notes 6** and **7**).
2. T7 RNA polymerase.
3. 5× transcription buffer: 200 mM Tris (pH 7.9), 30 mM $MgCl_2$, 10 mM spermidine, 50 mM NaCl.
4. 0.5 M DTT.
5. RNase-free H_2O.
6. 10 mM ATP, CTP, GTP, UTP.
7. 5-(3-aminoallyl)-UTP.
8. RNase-free DNase I.
9. Acid phenol-$CHCl_3$.
10. Ethanol.
11. Activated ester derivative of fluorescent dye (e.g., tetrafluorophenyl ester), dissolved in DMSO.
12. $NaHCO_3$.
13. DMSO.

3 Methods

3.1 3′ End Addition of Multiple Dyes to RNA Using Klenow

The ability of bacterial DNA polymerase I to use RNA as a primer for DNA synthesis led Huang and Szostak [14] to develop a method for labeling the 3′ end of an RNA of known sequence with a single radiolabeled nucleotide. Many DNA polymerases, including the Klenow fragment of DNA polymerase I, can efficiently incorporate nucleotides that carry modifications at specific nucleobase positions, in particular positions 4 and 5 of pyrimidines [15–18]. Flexible linkers that minimize steric hindrance support the incorporation of multiple bulky modifications, such as biotin, fluorescent dyes, or reactive groups (e.g., primary amine). Modified dUTP and dCTP derivatives are commercially available. We have successfully labeled RNA with a ~150-nt 3′ extension containing as many as 17 molecules of a fluorescent dye [19].

Sequence-dependent labeling of RNA with multiple dyes and biotin provides several unique benefits when used in CoSMoS and other single-molecule fluorescence experiments. First, multiple molecules of a fluorescent dye increase the signal intensity and reduce the laser power needed for imaging, which in turn minimizes photobleaching and extends observation time. Second, a multiply labeled RNA exhibits a gradual, step-wise photobleaching pattern that is distinct from events that cause an all-or-none change in fluorescence—e.g., loss of signal from dissociation or enzymatic cleavage of the RNA [19]. By contrast, photobleaching and all-or-none changes in fluorescence cannot be distinguished when an RNA is labeled with one molecule of a dye. Lastly, labeling by 3′ extension provides control over the order in which two different modifications (e.g., a biotin and a fluorescent dye) are added to an RNA. For example, 3′ extension ensures that every RNA with a 3′-terminal biotinylated nucleotide is also co-labeled internally with fluorescent dye. By contrast, labeling by chemical conjugation produces a sizable fraction of RNAs that are incompletely labeled. In CoSMoS experiments, a 3′-biotinylated RNA that lacks fluorescent dye will bind to the streptavidin-coated slide and it will recruit spliceosomal proteins, but the binding of spliceosomal protein to non-fluorescent RNA cannot be distinguished from nonspecific binding to the slide surface.

For the 3′-extension procedure, a DNA oligonucleotide must be designed as a template for Klenow (Fig. 1; *see* **Note 1**). The 3′ end of the oligonucleotide should be complementary to the last 15–20 nucleotides at the 3′ end of the RNA to be labeled, and the remaining sequences should be the complement of the labeled sequence extension. The RNA-DNA duplex formed between the 3′ ends of the RNA and DNA template must have a melting temperature (T_m) of at least 50 °C to ensure its stability at 37 °C during the extension reaction. Figure 1 shows an example where a DNA template is designed to make a 21-nt DNA extension labeled with two Alexa488 dyes followed by a single biotin. dUTP with Alexa488 dye attached to the position 5 of the uracil via a 5-atom linker (Alexa488-5-dUTP) and dCTP with biotin attached to the position 5 of cytosine via a 16-atom linker (biotin-16-dCTP) are used to incorporate the dye and biotin in positions where the template oligonucleotide has A and G, respectively. The template also encodes 6-nt stretches of adenosines to separate the modified bases and reduce interactions between the dyes, biotin-streptavidin complex, and the RNA. The template oligonucleotide and RNA are first denatured by heating up to 75 °C, and then annealed by slowly cooling to room temperature. Magnesium-free buffer is used to avoid Mg^{2+}-mediated RNA scission at elevated temperatures. The DNA extension is synthesized by Klenow polymerase in the presence of modified dNTPs. As an optional step, the DNA template can be removed from the labeled product by the addition of a large excess of oligonucleotide complementary to the DNA template, and briefly heating the mixture to 95 °C.

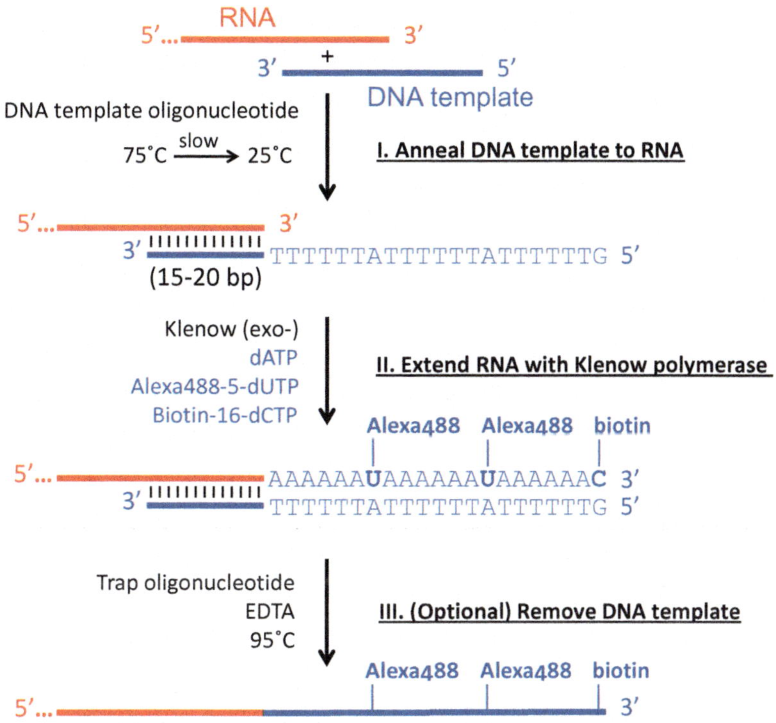

Fig. 1 3′ end labeling of RNA via template-directed DNA synthesis by the Klenow fragment of *E. coli* DNA polymerase I. In this example, two Alexa 488 dyes and a single biotin are incorporated into a 21-nt DNA (*blue*) extension at the 3′ end of the RNA (*red*)

1. Assemble the annealing reaction:

   50–500 pmol RNA.

   1.5-fold molar excess of oligonucleotide template.

   2 μL of 10× annealing buffer (Subheading 2.1).

   RNase-free water to a final volume of 20 μL.

2. Place the tube in a metal block preheated to 75 °C. Place the block on the bench and allow the reaction to slowly cool to room temperature over ~15 min. Alternatively, set up a PCR program that reduces the temperature by 5° every 2 min (75–25 °C, in 5° increments).

3. Assemble the labeling reaction:

   8 μL NEBuffer 2.

   0.5 mM dATP.

   0.1 mM Alexa488-5-dUTP (*see* **Note 2**).

   0.1 mM biotin-16-dCTP.

   20 μL annealing reaction (from **step 2**).

2 μL (10 units) of Klenow polymerase (exo-).

RNase-free water to a final volume of 80 μL.

4. Incubate the reaction at 37 °C for 1 h.
5. Add 4 μL 0.5 M EDTA to stop the reaction.
6. Add 8 μL 5 M ammonium acetate, 20 μg glycogen (optional; helps RNA recovery), and RNase-free water to a final volume of 100 μL. Precipitate RNA with 300 μL of ethanol, and resuspend the pellet in a small volume of RNase-free water (*see* **Note 3**).
7. (Optional DNA template removal step; *see* **Note 4**.) To the labeled RNA, add five- to ten-fold molar excess of DNA trap oligonucleotide. Heat for 1 min at 95 °C. Purify the labeled RNA on a denaturing gel (*see* **Note 5**).

3.2 Multi-dye Body-Labeled RNA Oligomer Preparation

Incorporation of multiple dyes at a specific position in a pre-mRNA reporter, e.g., in its intron allows more complex experiments than with a location reporter. Such an intron-labeled pre-mRNA reporter has been critical for a recent study of alternative spliceosome assembly pathways by Shcherbakova and colleagues [10]. Incorporation of multiple dyes yields bright RNA reporters and allows distinguishing intron loss by splicing from photobleaching as discussed above. Here, we describe how to prepare such a multi-dye body-labeled RNA oligomer. To obtain the full-length pre-mRNA reporter the dye-labeled RNA oligomer is subsequently ligated together with additional RNA oligomers as described previously in full detail [13] (*see* **Note 8**).

1. Assemble in vitro transcription reaction at room temperature replacing. A part of the UTP is replaced by 5-(3-aminoallyl)-UTP (typically 50%, but can be adjusted depending on the length and sequence composition and desired labeling efficiency):

   0.5 pmol of linear DNA template.

   RNase-free H_2O up to a final volume of 50 μL.

   2.5 μL of 0.5 M DTT.

   2 μL of 10 mM ATP.

   2 μL of 10 mM CTP.

   2 μL of 10 mM GTP.

   2-x μL of 10 mM UTP.

   x μL of 10 mM 5-(3-aminoallyl)-UTP.

   10 μL of 5× transcription buffer.

   2.5 μL of 20 U/μL T7 RNA polymerase.

2. Incubate reaction mix for 3 h at 37 °C.
3. Add 4 μL of 1 U/μL RNase-free DNase I and incubate for 20 additional minutes at 37 °C.

4. Extract reaction mix with acid phenol-$CHCl_3$.
5. Ethanol precipitate RNA and resuspend in 50 μL RNase-free H_2O.
6. Label RNA with an excess of the desired fluorescent dye-conjugate in a 100 μL reaction mix supplemented with fresh $NaHCO_3$ buffer (final concentration 0.1 M) for 3 h at room temperature in the dark with gentle agitation.
7. Resuspend the RNA in RNase-free water. RNA and dye concentration are assessed spectrophotometrically to determine the average number dyes per RNA molecule. The RNA concentration typically has to be determined from the absorbance at 260 nm corrected for absorbance due to the dye at this wavelength.

4 Notes

1. Klenow could use the RNA as a template and extend the DNA oligonucleotide, introducing aberrant labeled products and depleting modified dNTPs. Extension of the DNA oligonucleotide template can be prevented by designing the oligonucleotide with a 3′-terminal dideoxy nucleotide or by including one to more additional 3′ nucleotides that will not base pair to the RNA. Alternatively, if only three nucleotides comprise the sequence of the labeled extension (e.g., pyrimidines and adenine but lacks guanine, as in Fig. 1), the DNA oligonucleotide could be designed such that the first nucleotide needed to extend the 3′ end of the DNA oligonucleotide corresponds to the missing dNTP.
2. Long extension products that incorporate multiple modified bases may require an increased concentration of modified dNTPs. In general, we recommend that the amount of the modified dNTP in the extension reaction be at least twice the theoretical amount needed to synthesize the full-length DNA extension, or 0.1 mM, whichever is greater.
3. A single ethanol precipitation may be insufficient to remove unincorporated dNTPs. The presence of unincorporated nucleotides will artificially increase the concentration of the product—measured by absorbance—and could interfere with downstream applications. Perform a second ethanol precipitation or size-exclusion chromatography, or gel-purify the labeled extension product to remove excess nucleotides.
4. Many applications benefit from the removal of the DNA template from the labeled product after the extension reaction. When added to eukaryotic cell extracts, for example, the RNA-DNA hybrid in the labeled product (Fig. 1) could be a substrate

of RNase H. If the extension is short, purification of the product by denaturing gel electrophoresis will completely remove the DNA template. Products with extensions longer than 20–30 nucleotides, however, may form duplexes that cannot be completely melted in a standard denaturing gel, resulting in smeared bands. To ensure complete removal of the DNA template, a "trap" DNA oligonucleotide can be used to sequester the template. The trap oligonucleotide sequence must be the exact complement of the DNA template used for Klenow extension. Add a tenfold molar excess of the trap oligonucleotide. Briefly heat the sample to 95 °C to completely denature the extension product and oligonucleotide template. The trap oligonucleotide will sequester the DNA template and prevent it from reannealing to the labeled extension product. A sufficient amount of EDTA must be added prior to heating to bind magnesium ions present in the Klenow reaction buffer. Finally, gel-purify the labeled extension product by denaturing gel electrophoresis.

5. The yield of labeled product in an extension reaction will depend on several factors, including the nature of modified dNTPs, the length of DNA extension, and the accessibility of the 3′ end of the RNA. For RNAs transcribed in vitro by T7 polymerase, the most important factor that limits the labeling efficiency is heterogeneity at the 3′ ends of RNAs due to non-templated addition of one or more nucleotides. As much as 70% of RNA transcribed by T7 polymerase can have non-templated extensions [20] many of which will not be complementary to the DNA template and will fail to prime DNA synthesis by Klenow, thereby reducing the yield of labeled extension products. A detailed analysis of approaches to reduce the 3′ heterogeneity caused by T7 transcription is beyond the scope of this chapter, but a number of strategies have been reported. 2′-*O*-methyl modifications at the last two positions of the in vitro transcription template significantly reduce the fraction of RNA transcripts with non-templated nucleotides [21]. Alternatively, the 3′ end of the RNA transcript can be precisely removed using self-cleaving ribozymes [20, 22–25], DNAzymes [26], or RNase H [27], which will leave a 3′ end with known sequence.
6. Addition of extra CG nucleotides upstream to the canonical T7 sequence can increase the yield of in vitro transcription products [28].
7. T7 RNA polymerase can add non-templated bases at the 3′ end and result in a heterogeneous population of RNA products (*see* **Note 5**).
8. The two options, RNA ligase 1 and RNA ligase 2 for splinted RNA ligation, are discussed in detail in [13]. Therefore, we

only want to point out what to consider when choosing between RNA ligase 1 and 2: RNA ligase 2 requires base complementarity of the two RNA fragments to the DNA splint at the site of ligation, so the ligation product defined by the DNA splint is strongly favored over other ligation events. If heterogeneous populations of RNA fragments are used for the ligation (*see* **Note 5**) and the precise sequence at the ligation site is not of concern, then RNA ligase 1 is expected to produce higher yields than RNA ligase 2 since it does not require full complementarity of the two RNA fragments with the DNA splint.

9. The aminosilane-PEG surface to passivate the glass slide (*see* [13] for details) is subject to hydrolysis in aqueous environment. We recommend rinsing the slides directly after PEGylation to remove unbound PEG and storing them dried at −80 °C. No adverse effect of the drying and freezing procedure has been observed yet. After the initial rinse, the slide surface is exposed to aqueous solution only immediately before the experiment. Incubation of slides with solutions of pH ranges that accelerate hydrolysis should be avoided.

10. An internal control to quantify nonspecific binding in CoSMoS experiments: CoSMoS data is processed to determine the fluorescence intensity at areas of interest (AOIs), typically defined by the location of the pre-mRNA reporters. Fluorescence signal from the molecules of interest at individual AOIs is indicative of colocalization. For an appropriate interpretation of the observed fluorescence signals it is critical to assess their specificity. Any molecule of interest will bind, in addition to the specific colocalization events, with a certain affinity nonspecifically to the slide surface. To quantitatively assess nonspecific binding, control AOIs are chosen at random locations in the field of view and processed in the exact same way for spot detection as the AOIs defined by the locations of the pre-mRNA reporter. This analysis allows quantitative assessment of nonspecific binding directly during the CoSMoS experiment and is typically highly preferable to separate control experiments. Control AOIs in combination with additional biological control conditions are thorough controls, especially valuable for CoSMoS experiments in the full complexity of cellular extracts. A detailed description of CoSMoS data analysis and interpretation can be found here [29].

11. Rastergrams to visualize CoSMoS data: After spot detection CoSMoS datasets contain information on colocalization of the molecule(s) of interest at each AOI. These datasets can be analyzed statistically, but patterns due to different populations or other features in the single-molecule traces can be difficult to identify without a visual overview of the data. Here, we point

out rastergrams for intuitive CoSMoS data visualization [10]. Each row corresponds to an individual AOI and the colocalization status for one (Fig. 2d in [10]) or multiple components (Fig. 4d in [10]) over time is shown. The AOIs can be randomly arranged or sorted/aligned by a multitude of properties such as time to first binding event, longest/shortest binding event, or number of total binding events.

Acknowledgments

Joerg E. Braun acknowledges funding from the Human Frontier Science Program (HFSP) LT000166/2013. Joerg E. Braun and Victor Serebrov are supported by funding to the laboratory of Melissa J. Moore: University of Massachusetts Medical School, Investigator of the Howard Hughes Medical Institute and NIH R01 GM50037.

References

1. Berget SM, Moore C, Sharp PA (1977) Spliced segments at the 5′ terminus of adenovirus 2 late mRNA. Proc Natl Acad Sci U S A 74(8):3171–3175
2. Chow LT, Gelinas RE, Broker TR, Roberts RJ (1977) An amazing sequence arrangement at the 5′ ends of adenovirus 2 messenger RNA. Cell 12(1):1–8
3. Walter NG, Huang CY, Manzo AJ, Sobhy MA (2008) Do-it-yourself guide: how to use the modern single-molecule toolkit. Nat Methods 5(6):475–489. doi:10.1038/nmeth.1215
4. Ha T (2014) Single-molecule methods leap ahead. Nat Methods 11(10):1015–1018. doi:10.1038/nmeth.3107
5. Crawford DJ, Hoskins AA, Friedman LJ, Gelles J, Moore MJ (2008) Visualizing the splicing of single pre-mRNA molecules in whole cell extract. RNA 14(1):170–179. doi:10.1261/rna.794808
6. Abelson J, Blanco M, Ditzler MA, Fuller F, Aravamudhan P, Wood M, Villa T, Ryan DE, Pleiss JA, Maeder C, Guthrie C, Walter NG (2010) Conformational dynamics of single pre-mRNA molecules during in vitro splicing. Nat Struct Mol Biol 17(4):504–512
7. Hoskins AA, Friedman LJ, Gallagher SS, Crawford DJ, Anderson EG, Wombacher R, Ramirez N, Cornish VW, Gelles J, Moore MJ (2011) Ordered and dynamic assembly of single spliceosomes. Science 331(6022):1289–1295. doi:10.1126/science.1198830
8. Crawford DJ, Hoskins AA, Friedman LJ, Gelles J, Moore MJ (2013) Single-molecule colocalization FRET evidence that spliceosome activation precedes stable approach of 5′ splice site and branch site. Proc Natl Acad Sci U S A 110(17):6783–6788. doi:10.1073/pnas.1219305110
9. Krishnan R, Blanco MR, Kahlscheuer ML, Abelson J, Guthrie C, Walter NG (2013) Biased Brownian ratcheting leads to pre-mRNA remodeling and capture prior to first-step splicing. Nat Struct Mol Biol 20(12):1450–1457. doi:10.1038/nsmb.2704
10. Shcherbakova I, Hoskins AA, Friedman LJ, Serebrov V, Correa IR Jr, Xu MQ, Gelles J, Moore MJ (2013) Alternative spliceosome assembly pathways revealed by single-molecule fluorescence microscopy. Cell Rep 5(1):151–165. doi:10.1016/j.celrep.2013.08.026
11. Hoskins AA, Rodgers ML, Friedman LJ, Gelles J, Moore MJ (2016) Single molecule analysis reveals reversible and irreversible steps during spliceosome activation. Elife 5. doi:10.7554/eLife.14166
12. Semlow DR, Blanco MR, Walter NG, Staley JP (2016) Spliceosomal DEAH-box ATPases remodel pre-mRNA to activate alternative splice sites. Cell 164(5):985–998. doi:10.1016/j.cell.2016.01.025
13. Anderson EG, Hoskins AA (2014) Single molecule approaches for studying spliceosome assembly and catalysis. Methods

Mol Biol 1126:217–241. doi:10.1007/978-1-62703-980-2_17
14. Huang Z, Szostak JW (1996) A simple method for 3′-labeling of RNA. Nucleic Acids Res 24(21):4360–4361
15. Anderson JP, Angerer B, Loeb LA (2005) Incorporation of reporter-labeled nucleotides by DNA polymerases. BioTechniques 38(2):257–264
16. Langer PR, Waldrop AA, Ward DC (1981) Enzymatic synthesis of biotin-labeled polynucleotides: novel nucleic acid affinity probes. Proc Natl Acad Sci U S A 78(11):6633–6637
17. Gebeyehu G, Rao PY, SooChan P, Simms DA, Klevan L (1987) Novel biotinylated nucleotide–analogs for labeling and colorimetric detection of DNA. Nucleic Acids Res 15(11):4513–4534
18. Boyle A, Perry-O'Keefe H (2001) Labeling and colorimetric detection of nonisotopic probes. Curr Protoc Mol Biol Chapter 3:Unit3 18. doi:10.1002/0471142727.mb0318s20
19. Salomon WE, Jolly SM, Moore MJ, Zamore PD, Serebrov V (2015) Single-molecule imaging reveals that Argonaute reshapes the binding properties of its nucleic acid guides. Cell 162(1):84–95. doi:10.1016/j.cell.2015.06.029
20. Ferre-D'Amare AR, Doudna JA (1996) Use of cis- and trans-ribozymes to remove 5′ and 3′ heterogeneities from milligrams of in vitro transcribed RNA. Nucleic Acids Res 24(5):977–978
21. Kao C, Zheng M, Rudisser S (1999) A simple and efficient method to reduce nontemplated nucleotide addition at the 3 terminus of RNAs transcribed by T7 RNA polymerase. RNA 5(9):1268–1272
22. Wichlacz A, Legiewicz M, Ciesiolka J (2004) Generating in vitro transcripts with homogenous 3′ ends using trans-acting antigenomic delta ribozyme. Nucleic Acids Res 32(3):e39. doi:10.1093/nar/gnh037
23. Schurer H, Lang K, Schuster J, Morl M (2002) A universal method to produce in vitro transcripts with homogeneous 3′ ends. Nucleic Acids Res 30(12):e56
24. Di Tomasso G, Salvail-Lacoste A, Bouvette J, Omichinski JG, Legault P (2014) Affinity purification of in vitro transcribed RNA with homogeneous ends using a 3′-ARiBo tag. Methods Enzymol 549:49–84. doi:10.1016/B978-0-12-801122-5.00003-9
25. Gao Y, Zhao Y (2014) Self-processing of ribozyme-flanked RNAs into guide RNAs in vitro and in vivo for CRISPR-mediated genome editing. J Integr Plant Biol 56(4):343–349. doi:10.1111/jipb.12152
26. Pyle AM, Chu VT, Jankowsky E, Boudvillain M (2000) Using DNAzymes to cut, process, and map RNA molecules for structural studies or modification. Methods Enzymol 317:140–146
27. Lapham J, Crothers DM (1996) RNase H cleavage for processing of in vitro transcribed RNA for NMR studies and RNA ligation. RNA 2(3):289–296
28. Baklanov MM, Golikova LN, Malygin EG (1996) Effect on DNA transcription of nucleotide sequences upstream to T7 promoter. Nucleic Acids Res 24(18):3659–3660
29. Friedman LJ, Gelles J (2015) Multi-wavelength single-molecule fluorescence analysis of transcription mechanisms. Methods 86:27–36. doi:10.1016/j.ymeth.2015.05.026

Chapter 4

In Vitro Modulation of Endogenous Alternative Splicing Using Splice-Switching Antisense Oligonucleotides

Jeong Eun Park and Luca Cartegni

Abstract

Regulation of alternative splicing can be harnessed by antisense-based compounds to control gene expression. Antisense-mediated splicing interference has become a valuable molecular tool to modulate endogenous alternative splicing patterns, to correct cryptic or aberrant splicing, to reduce gene expression by triggering nonsense-mediated mRNA decay, and to activate intronic polyadenylation, both in vitro and in vivo. Here, we describe methods to induce and analyze the modulation of RNA processing, using modified splice-switching antisense oligonucleotides, such as phosphorodiamidate morpholino (PMO).

Key words mRNA, Antisense-oligonucleotide, Splicing, Polyadenylation, Morpholino

1 Introduction and Experimental Design

The use of antisense oligonucleotides (ASOs) is a well-established method for the regulation of gene expression through multiple mechanisms, including RNase H-mediated RNA degradation, inhibition of translation, and modulation of splicing [1, 2]. In addition, modified splice-switching ASOs (SS-ASOs) can also be applied to the activation of individual intronic polyA (IPA) sites, to specifically induce stable truncated splicing variants [3, 4]. Unlike classical antisense approaches, the primary goal of splice-switching ASOs is not to knock down the target RNA via its degradation or translational interference, but rather to affect the primary sequence of the expressed polypeptide, and/or to control the relative abundance of the splicing variants produced. Because the pursued effect here in not simply *quantitative* (the reduction of target mRNA expression), but rather *qualitative* (the switch to the expression of a product with different biological properties), in addition to the modifications to improve stability and performance in vitro and in vivo common to other antisense approaches, SS-ASOs must also

Yongsheng Shi (ed.), *mRNA Processing: Methods and Protocols*, Methods in Molecular Biology, vol. 1648,
DOI 10.1007/978-1-4939-7204-3_4, © Springer Science+Business Media LLC 2017

include chemical modification to ensure that mRNA-degrading activities, such as RNAse H, are not triggered.

Together, such modifications include changes in the oligonucleotide backbone, such as phosphorothioated (PS) backbones, phosphorodiamidated morpholino oligonucleotides (PMOs) and peptide nucleic acids (PNAs), as well as modifications of the ribose moiety with 2′-O-Alkyl group, (2′-O-methyl [2′-OME] or 2′-O-methoxyethyl [2′-MOE]), 2′Fluoro (2′F) and Locked Nucleic Acid (LNA) [1, 2]. These modifications contribute to impede RNase H-dependent cleavage, while enhancing nuclease resistance, binding specificity and efficiency, and cellular uptake [2]. In general, in order to modulate pre-mRNA splicing, SS-ASOs must interfere with pre-mRNA sequences involved in specific splice-site selection (Fig. 1a), recognized by the highly conserved splicing machinery and their regulatory factors [5–7]. Typically, the targeted sequences include either the 5′ splice site (5′ss, consensus AG/GURAGU), the branch point site where the lariat forms (BPS, consensus YUNAY) or the 3′ splice site (3′ss, consensus YAG/N, and poly-pyrimidine tract, PPT) of the exon to be suppressed, or any of the additional regulatory *cis*-elements which might affect the usage of a specific splice site [5]. These include both intronic and exonic enhancers and silencers, which respectively stimulate or repress it [5]. The location of such regulatory elements can be determined by a broad range of approaches, including computational analysis [8–13] looking for the consensus-binding motifs of known splicing factors, or experimental strategies such as functional SELEX, RNA-CLIP, and other approaches [6, 10, 13–16], deletion/mutation analysis of pre-mRNAs using minigene systems [7, 17–19], or oligonucleotide tiling approaches [20, 21].

In general, targeting the splice sites directly will inhibit their usage and—depending on the context—lead to one or a more of the following events (Fig. 1b): exon skipping (especially when the splice site is an essential determinant of exon identity), activation of a known alternative splice-site competing with it, activation of a nearby cryptic splice site (upstream or downstream), intron retention (particularly when very short introns are involved), or IPA activation (typically limited to 5′ss targeting). Targeting splicing enhancers (exonic or intronic) would lead to a similar range of outcomes, although the specific ones might be different, because a given splicing enhancer might affect the usage of multiple splice sites, both natural and cryptic, in its vicinity. Whereas the negative splicing interference approaches described above are conceptually simple (even if the outcome could be a complex resulting splicing pattern), positive regulation—the net activation of a splice site—is more challenging and usually requires some knowledge of the molecular regulatory elements at play. The simplest way is to target exonic or intronic silencers, if they can be mapped either computationally or experimentally (Fig. 1c.1). This is not always effective,

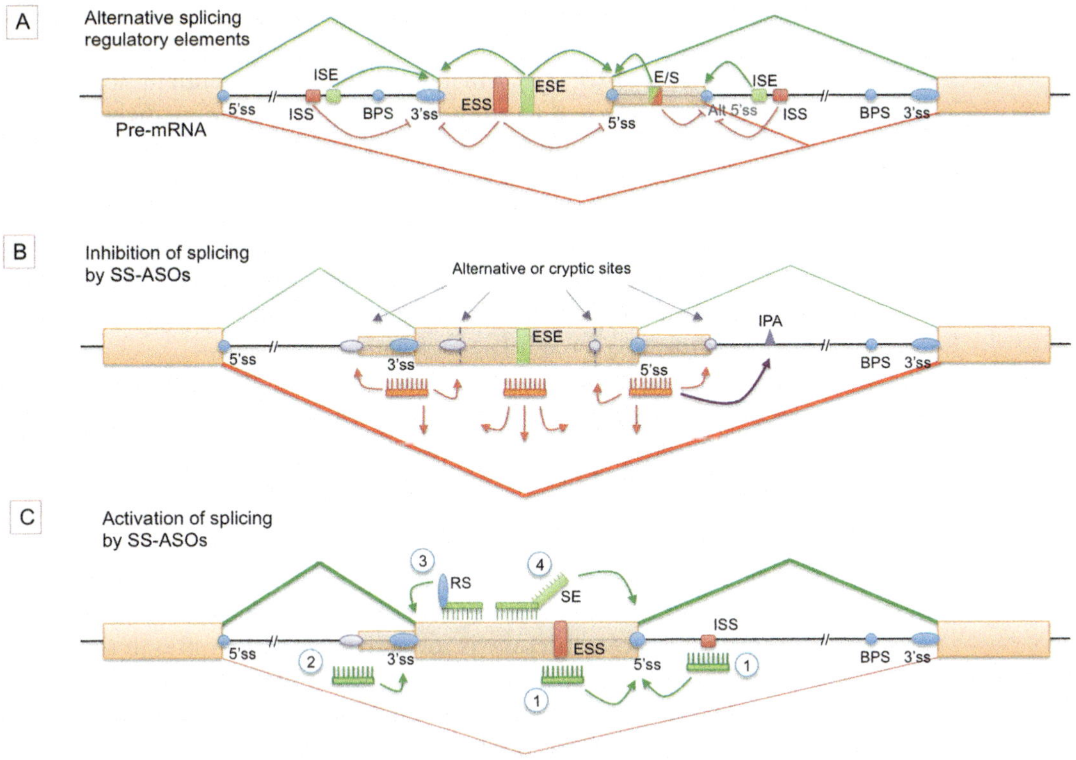

Fig. 1 Effects of Splice-Switching Antisense Oligonucleotides (SS-ASOs). (**a**) Splice site selection is controlled by multiple competing activating and inhibiting regulatory elements, located both in exons and introns. *ESE* = Exonic Splicing Enhancers, *ESS* = Exonic Splicing Silencers, *ISE* = Intronic Splicing Enhancers, *ISS* = Intronic Splicing Silencers, *BPS* = Branch Point Signal, 3′ss = 3′ splice site, 5′ss = 5′ splice site. In some cases, some regulatory elements can work as enhancers or silencers depending on position or context (E/S). (**b**) SSOs usually inhibit splicing when targeted to the splice sites or to enhancers. The most common outcome is skipping on the entire targeted exon. However, when alternative splice sites or cryptic splice sites are present within or around the exon, they may be activated at the expenses of the "normal" splice sites. When the 5′ss is targeted, Intronic PolyAdenylation (IPA) may be also triggered, if an actionable PolyA Signal (PAS) is present in the downstream intron (purple arrow). (**c**) To promote usage of a specific splice site, multiple strategies can be used, including (*1*) targeting exonic and intronic silencers, (*2*) targeting competing alternative splice sites, if present, (*3*) using ESSENCE compounds that mimic the action of stimulating splicing factors [22], or (*4*) using bifunctional ASOs that recruit such splicing factors to the exon [23]

as splicing silencers are often found in tandem with enhancers (especially in information-packed exons) and an SS-ASO that only affects (exonic) silencers might often be difficult to design. Intronic silencers tend to be more isolated, and thus represent more desirable targets. In the absence of silencer location data, enhanced splice site usage could be achieved by the negative targeting of a competing splice site (e.g., targeting an upstream 3′ss could improve inclusion of the downstream one, Fig. 1c.2) [18], or by the unbiased empirical testing of a large number of SS-ASOs tiling the relevant region (the major limit of this approach being its significant cost) [20, 21]. Finally, sequence-independent approaches have been proposed to stimulate splice site usage, namely the coupling of RS-rich peptide to generate compounds (named ESSENCE compounds, for exon-specific splicing enhancement by small chimeric effectors) that emulate the action of SR-proteins splicing factors and recruit the splicing machinery to the target site (Fig. 1c.3) [22], or the usage of bifunctional ASOs (Fig. 1c.4), comprising a targeting moiety and a recruiting moiety that contains a high affinity binding site for splicing factors [23].

The emphasis in this chapter is the splicing redirection that is achievable by ASO manipulation through direct targeting of splice sites, at the boundary between exon and intron (Fig. 1b). The immediate effect of targeting one of the splice sites varies from case to case, but—in general—it includes activation of a competing alternative splice site, induction of exon skipping, activation of putative cryptic splice sites in the vicinity, and, more rarely, induction of intron retention. Furthermore, when the 5′ss is targeted, another possible outcome is the activation of IPA sites in the intron downstream, because U1-snRNP, which recognizes the 5′ss and is essential for splicing, plays also an all-important additional role as a direct suppressor of downstream polyadenylation [4]. Therefore, blocking U1 snRNP binding at the 5′ss also releases its suppression of putative intronic polyA sites, which are very common. Indeed, intronic polyadenylation is an essential part of mRNA processing, and 50–60% of human genes contain alternative polyA sites (PASs) involved in alternative splicing [24–26]. IPA can generate shortened mature mRNAs and thus express truncated isoforms with radically altered biology, including possible dominant-negative capabilities. When designing splice-switching experiments, it is essential to recognize that the various possible outcomes are not mutual exclusive, but they could occur simultaneously, giving rise to complex splicing patterns.

A further complicating situation is the putative triggering of mRNA Nonsense-Mediated Decay (NMD) by many of the splicing events. The NMD pathway is a translation-dependent, conserved mRNA surveillance mechanism in eukaryotic cells that detects and specifically degrades mRNAs harboring premature termination codons (PTCs), to prevent the expression of putative deleterious variants [27]. Most alternative splicing events (natural or induced by SS-ASOs) will cause either an insertion or a deletion in the open reading frame (ORF) of the target gene. If the change is associated with a frameshift (which statistically will occur in about 2/3 of cases), then likely a PTC will be encountered in the new frame at some distance downstream of the frameshift, thus typically triggering NMD. Because such NMD-triggering events are associated with destabilized mRNAs, their occurrence would be vastly underestimated—and occasionally barely detectable—by standard PCR-based analysis of endogenous modulated splicing, which looks at steady-state mRNA levels. The possible induction of such variants should be thus taken into consideration when designing and analyzing splicing interference experiments. To better establish the true extent by which an SS-ASO treatment stimulates NMD-substrate variants, NMD can be inhibited, or mini-gene transfection experiments where no translation occurs can be performed.

The standard approach to analyze the effects of SS-ASOs on splicing is reverse transcription-polymerase chain reaction (RT-PCR), which is one of the most convenient methods for studying RNA transcripts. To account for the possible multiple outcomes of the SS-ASO treatment, it is advisable, when possible, to design PCR primers that span multiple exons upstream and downstream, with the splice site targeted by the SS-ASOs at the center, to ensure that multiple outcomes (exon skipping/inclusion, alternative 3′ or 5′ss, possibly multiple exon skipping) are detected (Fig. 2). As mentioned, when 5′ss are targeted, an additional possible outcome is the activation of IPA. Such products do not encompass exonic regions at its 3′ end, and thus would not be detected by the normal RT-PCR described above. In order to verify that IPA is or is not activated, a dedicate PCR should be run with the reverse primer located in the regions of the intron immediately downstream of the 5′ss, upstream of the predicted IPA site (primer Rev_2, Fig. 2a). When assessing exon/inclusion skipping or usage of alternative splice sites usage, a simple low cycle RT-PCR using the external exonic primers is sufficient for semiquantitative results, and certainly to get a good indication of the effect of a given SS-ASO. While using low-cycle radioactive PCR [17] increases the accuracy of quantitation, the results remain essentially the same when standard nonradioactive RT-PCR is used, especially if care is taken to use a lower number of cycles, so that the amplification reaction has not reached plateau yet, but it is still in its exponential phase. We suggest a real-time RT PCR to be run initially, to determine the ideal number of cycles for a given PCR. However, when IPA is involved, low-cycle three-primer PCR should be used instead. This involves the simultaneous usage of two downstream primers, one exonic and one intronic, as illustrated in Fig. 2. The two PCR products are co-amplified and their ratio gives a semiquantitative measure of the relative template abundance (and thus of the effects of the treatment). In order to obtain reliable results, it is essential that the competing primers be optimized and calibrated, for comparable amplification efficiencies. This should be done at the onset, using multiple primer sets for each variant. The calibration can be done measuring efficiency by real-time PCR using known quantities of subcloned templates or, (recommended) by performing an "inverted templates curve" with the two expected products (Fig. 2b, right panel). In this case, the expected product

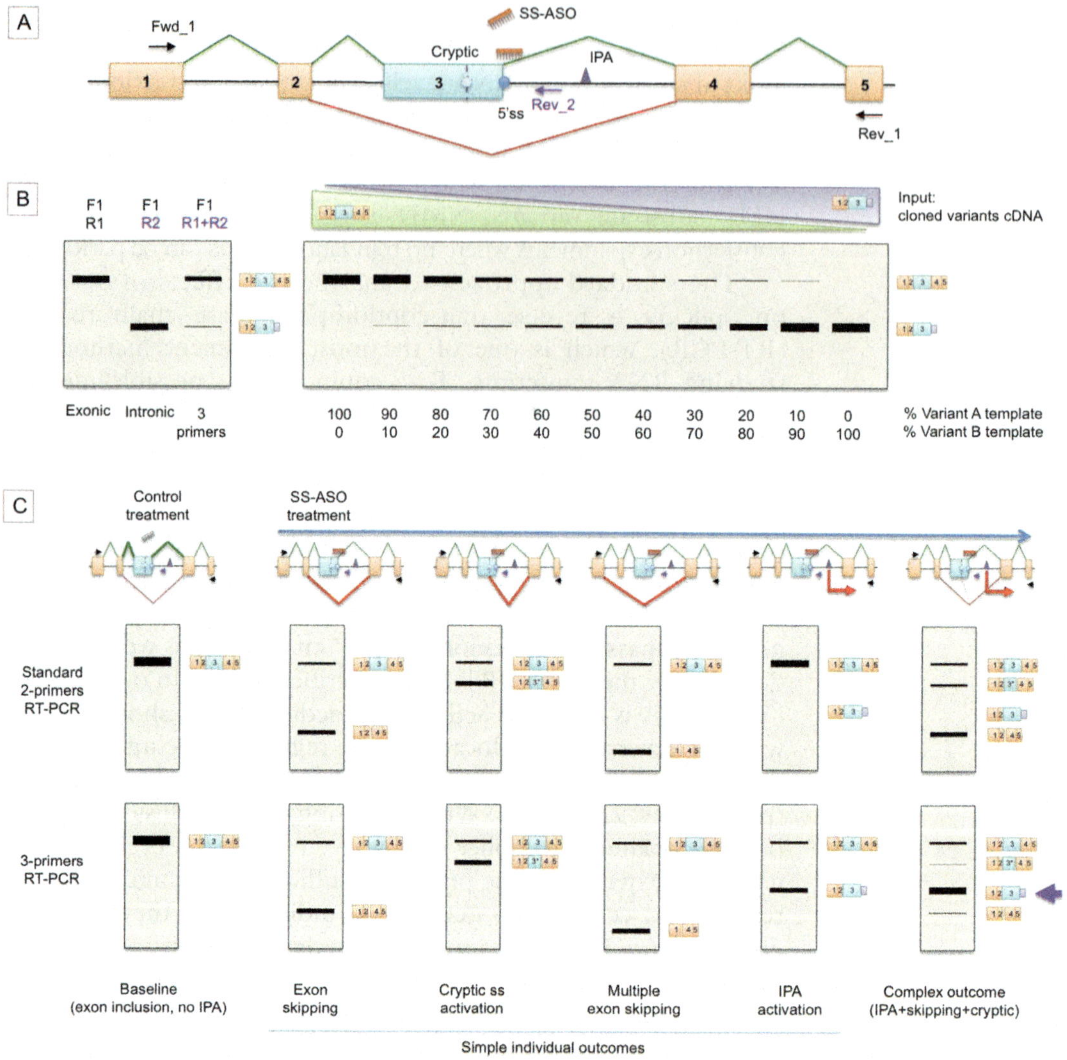

Fig. 2 Analysis of the effects of SS-ASOs targeted to the 5′ss. (**a**) Schematic of a typical pre-mRNA substrate, with some of the possible common alternative splicing events. Approximate positioning of PCR primers for 2- and 3-primer PCR is shown. If possible, the exonic primers should be positioned, at least one or two exons upstream or downstream of the target splice site. The reverse intronic primer should be positioned shortly downstream of the splice site, and upstream of any putative PAS. (**b**) Results and calibration of 3-primer PCR. Initially, two separate RT-PCRs are run with a common forward primer and the two separate reverse primers (exonic and intronic). A 3-primer PCR will yield two bands if IPA is involved. Because two separate PCRs are run simultaneously in competing conditions, an inverted calibrating curve is performed using variable amount of recombinant substrates to ensure that the observed product ratios correspond to the template input ratios. (**c**) Possible effects of SS-ASOs targeted to the 5′ss, when analyzed by 2- (*top series*) or 3-primer PCR (*bottom series*). Hypothetical results of an SS-ASO targeted to the 5′ss are shown left to right. On the left the baseline of untreated cells is shown. Some of the possible common outcomes are exon skipping, activation of a cryptic site (exonic as shown, or intronic, not shown), skipping of multiple exons, and activation of IPA. The last panels on the right show the possible coexistence of several of these events. In this hypothetical situation, a simple 2-primer PCR (*top*) would lead to the conclusion that the major product is exon 3 skipping with some cryptic activation, whereas the more accurate 3-primer PCR reveals that IPA activation is the principal outcome (*purple arrow*), with the latter two events only occurring at trace levels

needs to be cloned as cDNAs and a set of 3-primer PCRs are run with different relative template inputs. PCR condition and/or primer set should be optimized so that the product ratio corresponds to the input. If this is not obtainable, data from the inverted template curve can be used as a standardization curve to infer the actual variants ratio. Appropriate calibration of PCR efficiency is essential when IPA is involved, as the two reverse primers could have significantly different efficiency, and/or the PCR products are different enough in sequence and size that their co-amplification profile could produce highly misleading results if not calibrated. Similarly, when targeting 5′ss, it is equally essential to test for possible effects on putative intronic IPA sites, even if IPA variants had not been previously described in the downstream intron. In our experience, actionable IPA sites are relatively common, and disregarding them could lead to significant data misinterpretation. Figure 2c represents various possible outcomes of an SS-ASO treatment targeting a hypothetical 5′ss site. Whereas standard 2-primer PCR gives reliable results when exon skipping or cryptic/alternative sites are involved, if IPA is at play, the standard PCR results would be highly misleading.

2 Materials

Solutions should be prepared with RNase-free and diethylpyrocarbonate (DEPC)-treated double-distilled water to prevent RNA degradation. All commercial reagents or chemicals must be ultrapure grade for molecular biology.

2.1 RNA Source

All experiments can be examined using total RNA and extracted from mammalian cells grown in tissue culture system.

2.2 SS-ASOs

As mentioned, multiple chemistries and delivery methods can be employed.

Here, we will focus on vivo-morpholinos (Mo-ASOs), which are phosphorodiamidated ASOs coupled to a delivery moiety, which is comprised of an octa-guanidine dendrimer (vivo-morpholino, Gene Tools http://www.gene-tools.com).

1. Delivery aspects. Vivo morpholino efficiently self-deliver to a broad range of cell types in vitro and can be delivered to multiple organs, including tumors, in vivo. Vivo-Morpholinos enter cells by associating with the cell membrane, followed by endocytosis and subsequent lysosomal escape. The dendrimer moiety can be toxic to cells, and different cell types can have significantly different sensitivity to the toxic effect of control (nonspecific) vivo morpholino, from as low as 1 μM to as high as 15–20 μM.

Similarly, other chemistry/delivery strategies are associated with variable toxicity, and it is thus essential to determine the treatable range of a given cell line using control SS-ASOs.

2. Design tips. Mo-ASOs are usually 18–25 m. There are some limitation in terms of base composition and primary sequence especially as far as the G content is involved, as G and G-runs limit solubility. Care should be used to avoid any RNA secondary structure and self-dimerization capabilities, as morpholino duplexes are very stable and such structures would effectively compete with target hetero-duplexing (same is true for PNAs and LNAs). Specific guiding design directions and more details for Mo-ASOs can be obtained at the Gene-Tools web site.
3. Targeting. If two alternative splice sites are in play, the competing splice site (3′ss or 5′ss) should be targeted first. Where IPA is desirable, then the 5′ss should be targeted. If the goal is to induce exon skipping and not to activate putative IPA sites, then the 3′ss is a much preferable target.

   To target a 5′ss, a good start is an ASO (e.g., ASO_−2) targeting the last 2–3 exonic nucleotides and the first 18–22 nucleotides of the intron (compatibly with the requirements/limitations of the chosen chemistry). If budget allows it (a single synthesis of vivo morpholino currently costs around $700), multiple staggered compounds should be tested, as their effectiveness can vary greatly. For example, in addition to the "−2" compound hypothesized above, a "ASO_−6" and a "ASO_+2" could be synthesizes and tested. To target a 3′ss, a similar strategy can be employed. However, because the Pyrimidine tract is often rich in cytidines, this could impede ASOs synthesis/efficacy (because of the deriving high G content). Therefore, ASOs targeting the 3′ss should often be skewed toward the inside of the exon.
4. Control ASOs. Common control ASOs include "scrambled" ASOs of the same length and base composition of the test compound, "inverted" ASOs comprising the same sequence as the test compound but in a 3′->5′ orientation (NOT the antiparallel, complementary sequence), or "mutated" ASOs, where multiple (2–4) single-base mismatches are introduced. The sequence of the control ASOs should be queried against databases (e.g., BLAST) to ensure that they are not likely to recognize and affect the processing of other transcripts. Whereas ideally each test compound would have its own paired control, this is highly unpractical in reality, and could become very expensive. Therefore, a single common "random" control is usually used as a control for any battery of test compounds.

2.3 RNA Extraction from Mammalian Cells

2.3.1 Tissue Cell Culture

1. Complete medium (GIBCO): 10% (v/v) fetal bovine serum (FBS), 2000 mg/ml D-glucose 300 mg/ml L-glutamine, 1% penicillin-streptomycin solution (10,000 units penicillin and 10 mg streptomycin/ml in 0.9% NaCl. Store at 4 °C.
2. Low serum medium containing 4% (v/v) FBS.
3. Phosphate-buffered saline (PBS): 1 mM (144.0 mg/l) KH_2PO_4, 155 mM (9000 mg/l) NaCl, 2.96 mM (795 mg/l) $Na_2HPO_4 \cdot 7H_2O$. Adjust to pH 7.4.

2.3.2 RNA Isolation and RT-PCR

1. β-mercaptoethanol (β-ME), or 2 M dithiothreitol (DTT).
2. 96–100% ethanol.
3. RNeasy Mini Kit (QIAGEN) with mini-spin columns, collection tubes (1.5 and 2 ml). Store at room temperature (15–25 °C).
4. SuperScript III First-Strand Synthesis System for RT-PCR (Invitrogen).
5. AccuPrime Taq DNA Polymerase System (Invitrogen).
6. Three PCR primers (One of forward primer and two of reverse primer sets) for specific splice isoform detection.

2.4 Agarose Gel and Electrophoresis

1. 0.5× Tris-borate-EDTA (TBE) buffer: 45 mM Tris, 45 mM boric acid, 1 mM EDTA for electrophoresis.
2. 2–2.5% (v/v) agarose gel.
3. 0.5 μg/ml Ethidium bromide.

2.5 Equipment

1. NanoDrop Spectrophotometer, such as DeNOVIX DS-11, for quantification of RNA.
2. A thermal cycler, such as the Bio-Rad DNA engine Dyad Peltier, for Reverse Transcriptase reaction and RT-PCR.
3. UV transilluminator with camera such as the Bio-Rad Gel Doc EZ Imager with Image Lab (Version 4.1) Analysis Software for visualization and densitometric analysis of ethidium bromide stained PCR product.
4. A tabletop centrifuge with cooling control such as SORVALL LEGEND MICRO 21R (Thermo Scientific).

3 Methods

3.1 Morpholino-ASOs Treatment

1. Seeded mammalian cells in 24-well tissue culture plates are grown to reach about 70–80% confluency in complete medium containing 10% (v/v) prior to Morpholino-ASOs treatment.
2. Vivo-morpholinos are added directly to cells seeded in 500 μl low serum medium containing 4% (v/v) FBS for 24–72 h (*see* **Note 1**).

It is advisable to perform a broad dose curve with the compounds, ranging from a few hundred nanomolar to low micromolar range, in parallel to control Mo-ASOs.

Most compounds will be effective around 2–8 μM. Active compounds will induce 50% splice switching at around 1–2 μM, but some can work as low as 100–200 nM. For those compounds whose activity becomes evident only above 4 μM, nonspecific toxicity (due to the dendrimer moiety) could become an issue, depending on the cell line used. Control Mo-ASOs need to be used to determine the upper usable limit for each cell line.

Depending on the reversibility of the biological effect, active SS-ASOs can be washed out by changing the media, and their molecular effect on splicing is fully reversed in 24–72 h.

3.2 RNA Isolation

1. Total RNA from mammalian cells treated with Vivo-morpholinos is extracted using the RNeasy Mini Kit (QIAGEN), following the product manual. For RNA extraction via this kit, cells can be directly lysed using the provided lysis buffer (*see* **Note 2**).
2. As described by the manufacturer-provided protocol, add one volume of 70% ethanol to the cell lysate and mix completely by pipetting. Then transfer the mixed sample to the spin column, and centrifuge at 8000 × *g* for 30 s. Discard the flow-through.
3. Wash the spin column once with 700 μl of RNA wash buffer RW1 and twice with 500 μl of RNA wash buffer RPE. Centrifuge at 8000 × *g* for 1 min between each wash step and discard the flow-through.
4. Place the spin column in a fresh 2 ml collection tube, and then remove the residual RNA RPE wash buffer from the column by additional centrifugation at full speed for 1 min.
5. Place the spin column into a new 1.5 ml microcentrifuge tube, then add 30 μl RNase-free water directly at the center of the spin column membrane.
6. Elute total RNA by centrifugation at full speed for 1 min, and determine RNA yield using a UV spectrometer NanoDrop (*see* **Note 2**).

3.3 RT Reaction for First Strand cDNA Synthesis

1 pg–5 μg of total RNA can be used to synthesize first-strand cDNA using the SuperScript III First-strand synthesis system (Invitrogen).

1. Proceed as described by the manufacturer-provided protocol: to a 10 μl reaction containing the total RNA, add 1 μl of 50 μM oligo $(dT)_{20}$ or 0.05 μg random hexamers and 1 μl of 10 mM dNTP. Mix gently and spin briefly.

2. Denature mixed RNA/oligo at 65 °C for 5 min, then place on ice for 1 min.
3. Add 10 μl of cDNA synthesis mix, containing 2 μl 10× RT buffer, 4 μl 25 mM $MgCl_2$, 2 μl 0.1 M DTT, 1 μl RNaseOUT™ (40 U/μl), and 1 μl SuperScript III RT (200 U/μl) to each RNA/oligo mixture. Mix gently and spin briefly.
4. Incubate the oligo$(dT)_{20}$-primed mixture at 50 °C for 50 min, or incubate the random hexamer-primed mixture at 25 °C for 10 min. Spin briefly and inactivate the RT enzyme at 85 °C for 5 min, then place on ice.
5. Add 1 μl of RNase H to cDNA synthesis reaction and incubate at 37 °C for 20 min. The cDNA product should be stored at −20 °C for long-term or used for PCR immediately (*see* **Note 3**).

3.4 Primer Design for Detection of Splicing Isoforms

As an example, we will use a 5-exons pre-mRNA where the 5′ss of exon 3 is targeted, and IPA is possibly induced (Fig. 2).

1. Following alternative splicing, possible splicing product includes the full-length mature mRNA, possible shorter isoforms generated by exon skipping, putative additional variants due to the usage of cryptic splice-sites and truncated variants activated by IPA, which include regions at the 5′ end of intron 3. Three compatible primers are designed to detect multiple splicing isoforms, with an ideal amplicon size of not more than 500 base pairs (*see* **Note 4**).
2. The common forward primer should be located at least one exon upstream of the targeted exon, to ensure that at least one splicing event is included in the final product. This guarantees that the product cannot derive from contaminating genomic DNA or from unspliced pre-mRNA (*see* Fig. 2).
3. Two different downstream reverse primers are designed for 3-primers RT-PCR analysis. One lies within the intronic span the downstream intron (exon 3 in the example in Fig. 2), positioned upstream of the first predicted PAS. The second is located on one of the next downstream exons (e.g., exon 5). The two exonic primers amplify the full-length product as well as exon skipping and cryptic/alternative sites variants. The intronic primer (paired to the upstream forward primer) is essential to detect the extent of IPA activation. The three oligonucleotides together measure the relative abundance of the alternative splicing variants in normal conditions or following SS-ASOs treatments.
4. In addition, standard PCR controls should be used, and the amplification of a uniformly expressed "house-keeping" gene (e.g., GAPDH), unaffected by the treatments, should be used for RNA normalization.

3.5 Performing and Analyzing RT-PCR to Detect Alternatively Spliced Isoforms

In vitro amplification of DNA by PCR performed using the AccuPrime *Taq* DNA Poymerase System (Invitrogen) by following the product manual.

1. In brief, in a 25 μl reaction, add 0.5–1.0 μl cDNA with 2.5 μl 10× PCR Buffer I, 0.5 μl *Taq* DNA polymerase, and 0.5 μl primer mix (10 μM each) (*see* **Note 5**). Mix and spin briefly. Overlay the reaction mixtures with one drop of mineral oil to prevent evaporation during the repeated heating and cooling cycles.
2. Complete PCR reaction using the following initial conditions for the thermal cycler: 95 °C pre-denaturation for 2 min, 25–35 cycles of PCR amplification of 95 °C for 15–30 s (Denaturing), 55 °C–60 °C for 15–30 s (Annealing), 68 °C for 1 min per kb (Extension), and lastly 68 °C for 10 min (*see* **Note 6**). Maintain the reaction at 4 °C after cycling. Stored at −20 °C until use.
3. The PCR products are separated by standard horizontal gel electrophoresis, on 2–2.5% gels, stained with 0.5 μg/ml ethidium bromide in 0.5× TBE, depending on the expected size of the amplified bands corresponding to the alternative splicing variants. A set of putative results is illustrated in Fig. 2c. Gels are imaged following UV transillumination using a Bio-Rad Gel Doc EZ Imager (or similar equipment) with Image Lab. Bands are then quantified with Image J quantitation software.

4 Notes

1. The complete medium containing 10% FBS is aspirated out, washed with PBS twice, and medium changed with fresh low serum medium containing 4% FBS before SS-ASO compound treatment.
2. RNeasy Mini Kit (QIAGEN) minimizes genomic DNA contamination. Follow manufacturer's instructions for the appropriate volumes of RLT lysis buffer. To increase RNA yield, repeat **step 6** or re-elute from **step 6**. An absorbance ratio between 1.8 and 2.1 of the 260 and 280 nm absorbance values (A260/280) indicates clean RNA.
3. All these incubations are best performed in a programmable thermocycler. For efficient and sensitive cDNA synthesis, RT reactions may be optimized from 1 pg to 5 μg of RNA. The sensitivity of the PCR step can be increased by removing the RNA template from the cDNA/RNA hybrid molecule by digestion with RNase H after first-strand synthesis. cDNA product should not exceed 1/10 of the PCR reaction total volume.

4. PCR amplicons for exon skipping splice isoform will be smaller than full length with normal splicing. Amplicons corresponding to cryptic/alternative sites can be either shorter or longer depending on the position of the alternative splice site. The third IPA-specific primer should be designed so that the resulting amplification product is easily distinguishable from the full-length and other alternative splicing products on a gel. It is very important that all amplicons are amplified with similar efficiency during the PCR reaction. Multiple PCR sets and conditions should be tested, in order to achieve this. Efficiency can be measured by individual 2-primers real-time PCRs, or by testing the primer combinations against different pairwise ratio mixtures of cloned cDNAs, corresponding to the variants of interest, as illustrated in Fig. 2b. Good primer design is essential. Matching melting temperatures (T_m) of the primers should be around 60 °C, primer length 18–23 nucleotides, 40–60% GC content. Multiple software options are available for PCR primer design, including free online tools such as the software OligoAnalyzer 3.1 (https://www.idtdna.com/calc/analyzer) by Integrative DNA Technologies to analyze T_m hairpin formation, self-dimerization, or hetero-dimerization, etc.
5. Prepare a PCR master mix containing *Taq* DNA polymerase for multiple reactions to minimize pipetting variations. Each set of PCRs should include control both a positive control (to monitor PCR efficiency) and a negative control (no template DNA) to detect DNA contaminations.
6. Each PCR cycle theoretically doubles the amplicon in the reaction. Primers, amplicon length, primary sequence, and base content all affect the actual efficiency. Different series of PCR conditions should be explored, to find the optimal conditions where the various isoforms amplify comparably [28]. The lowest possible number of cycles, at which the product is still detectable should be employed, for more accurate quantification. This information can be obtained by running a real-time PCR under the same conditions. For added accuracy, radioactive low-cycle PCR should be performed [17].

References

1. Spraggon L, Cartegni L (2013) Antisense modulation of RNA processing as a therapeutic approach in cancer therapy. Drug Discov Today Ther Strateg 10(3):e139–e148
2. Kole R, Krainer AR, Altman S (2012) RNA therapeutics: beyond RNA interference and antisense oligonucleotides. Nat Rev Drug Discov 11(2):125–140
3. Vorlova S et al (2011) Induction of antagonistic soluble decoy receptor tyrosine kinases by intronic polyA activation. Mol Cell 43(6):927–939
4. Spraggon L, Cartegni L (2013) U1 snRNP-dependent suppression of polyadenylation: physiological role and therapeutic opportunities in cancer. Int J Cell Biol 2013:846510
5. Cartegni L, Chew SL, Krainer AR (2002) Listening to silence and understanding nonsense: exonic mutations that affect splicing. Nat Rev Genet 3(4):285–298

6. Liu HX, Zhang M, Krainer AR (1998) Identification of functional exonic splicing enhancer motifs recognized by individual SR proteins. Genes Dev 12(13):1998–2012
7. Schaal TD, Maniatis T (1999) Multiple distinct splicing enhancers in the protein-coding sequences of a constitutively spliced pre-mRNA. Mol Cell Biol 19(1):261–273
8. Fairbrother WG et al (2002) Predictive identification of exonic splicing enhancers in human genes. Science 297(5583):1007–1013
9. Cartegni L et al (2003) ESEfinder: a web resource to identify exonic splicing enhancers. Nucleic Acids Res 31(13):3568–3571
10. Van Nostrand EL, Huelga SC, Yeo GW (2016) Experimental and computational considerations in the study of RNA-binding protein-RNA interactions. Adv Exp Med Biol 907:1–28
11. Zhang XH, Chasin LA (2004) Computational definition of sequence motifs governing constitutive exon splicing. Genes Dev 18(11):1241–1250
12. Aznarez I et al (2008) A systematic analysis of intronic sequences downstream of 5′ splice sites reveals a widespread role for U-rich motifs and TIA1/TIAL1 proteins in alternative splicing regulation. Genome Res 18(8):1247–1258
13. Wang Y, Wang Z (2014) Systematical identification of splicing regulatory cis-elements and cognate trans-factors. Methods 65(3):350–358
14. Liu HX et al (2000) Exonic splicing enhancer motif recognized by human SC35 under splicing conditions. Mol Cell Biol 20(3):1063–1071
15. Tian H, Kole R (2001) Strong RNA splicing enhancers identified by a modified method of cycled selection interact with SR protein. J Biol Chem 276(36):33833–33839
16. Wang Z et al (2004) Systematic identification and analysis of exonic splicing silencers. Cell 119(6):831–845
17. Cartegni L et al (2006) Determinants of exon 7 splicing in the spinal muscular atrophy genes, SMN1 and SMN2. Am J Hum Genet 78(1):63–77
18. Zammarchi F et al (2011) Antitumorigenic potential of STAT3 alternative splicing modulation. Proc Natl Acad Sci U S A 108(43):17779–17784
19. Lefave CV et al (2011) Splicing factor hnRNPH drives an oncogenic splicing switch in gliomas. EMBO J 30(19):4084–4097
20. Hua Y et al (2007) Enhancement of SMN2 exon 7 inclusion by antisense oligonucleotides targeting the exon. PLoS Biol 5(4):e73
21. Hinrich AJ et al (2016) Therapeutic correction of ApoER2 splicing in Alzheimer's disease mice using antisense oligonucleotides. EMBO Mol Med 8(4):328–345
22. Cartegni L, Krainer AR (2003) Correction of disease-associated exon skipping by synthetic exon-specific activators. Nat Struct Biol 10(2):120–125
23. Ghigna C et al (2010) Pro-metastatic splicing of Ron proto-oncogene mRNA can be reversed: therapeutic potential of bifunctional oligonucleotides and indole derivatives. RNA Biol 7(4):495–503
24. Hoque M, Li W, Tian B (2014) Accurate mapping of cleavage and polyadenylation sites by 3′ region extraction and deep sequencing. Methods Mol Biol 1125:119–129
25. Tian B et al (2005) A large-scale analysis of mRNA polyadenylation of human and mouse genes. Nucleic Acids Res 33(1):201–212
26. Hoque M et al (2013) Analysis of alternative cleavage and polyadenylation by 3′ region extraction and deep sequencing. Nat Methods 10(2):133–139
27. Popp MW, Maquat LE (2013) Organizing principles of mammalian nonsense-mediated mRNA decay. Annu Rev Genet 47:139–165
28. Horner RM (2006) Relative RT-PCR: determining the linear range of amplification and optimizing the primers:competimers ratio. CSH Protoc 2006(1)

Chapter 5

Purification of mRNA Processing Complexes Using an RNA Affinity Approach

Xiuye Wang and Yongsheng Shi

Abstract

Multiple mRNA processing steps, including splicing and 3′ processing, take place in macromolecular complexes that contain many proteins and sometimes RNA molecules. A key challenge in the mRNA processing field has been to define the structure-function relationship of these sophisticated molecular machines. A prerequisite for addressing this challenge is to develop tools for purifying mRNA processing complexes in their native and intact forms that are suitable for functional and structural studies. Among many methods that have been developed, RNA affinity-based methods are most widely applied. In these methods, RNA molecules that are substrates to mRNA processing machineries are fused with an affinity tag, incubated with cellular extracts/lysates to allow for the assembly of mRNA processing complexes, and finally the assembled complexes are purified using RNA affinity tag. In this chapter, we will overview RNA affinity-based purification methods and describe in detail one such method, MS2-tagging, and its application in the purification of mRNA 3′ processing complexes. Although these methods were originally developed for purifying mRNA processing complexes, they should be applicable to purification of other RNA-protein complexes as well.

Key words mRNA processing complex, Affinity purification, MS2

1 Introduction

mRNA processing takes place in several large macromolecular complexes. The mammalian spliceosomes consist of five small nuclear RNAs (snRNAs) and over one hundred proteins [1]. Mammalian 3′ processing complex contains over 20 core components and ~65 associated proteins [2]. In addition, these large molecular machines are highly dynamic. For example, the assembly of mammalian spliceosomes goes through multiple stages, from E, A, B, to C complexes, in which splicing reactions take place [1, 3]. The complexity and dynamic nature of mRNA processing complexes present great challenges to the study of their structure-function relationships.

A number of methods have been developed for the purification of mRNA processing complexes and they can be classified into two

Yongsheng Shi (ed.), *mRNA Processing: Methods and Protocols*, Methods in Molecular Biology, vol. 1648,
DOI 10.1007/978-1-4939-7204-3_5, © Springer Science+Business Media LLC 2017

general types. The first type is protein affinity-based methods. For example, immunoprecipitation of a number of epitope-tagged yeast splicing factors has recently been used to purify spliceosomes at different stages [4–6]. The advantages of this method include its simplicity and the ability to elute the tagged proteins under native conditions. In fact, spliceosomes purified using this method have been successfully studied by cryo-electron microscopy (cryo-EM) [7], providing near atomic resolution structures for the first time. The main disadvantage of this method, however, is the lack of specificity: all tagged proteins, including free proteins and those assembled into mRNA processing complexes, are purified in a complicated mixture. To achieve the purity required for some biochemical or structural studies, additional purifications steps are usually needed.

The second type of method for purifying mRNA processing complexes is based on RNA affinity. In this method, an RNA substrate for a specific mRNA processing machinery is fused with an affinity tag and then incubated with cell lysates/extract to allow for the assembly of the proper complex. The mRNP complexes are then purified using immobilized proteins or small molecules that specifically bind to the RNA affinity tag. Many RNA affinity tags have been introduced for this purpose, including biotin [8], artificially selected aptamers [9, 10], and several hairpin structures from bacteriophages [11]. Biotinylated RNA substrates have long been used to various mRNA processing complexes [12, 13]. This method takes advantage of the avid and specific interaction between biotin and streptavidin and allows purification under stringent conditions with a high level of recovery. However, biotin-based affinity purification also has its drawbacks: biotin labeling of RNA is not 100% efficient and, as a result, a significant portion of the RNA substrates is unlabeled and thus needs to be removed before purification. More importantly, it is very difficult to elute the purified mRNP complexes in its intact form for subsequent functional and/or structural studies. On the other hand, aptamer or natural hairpin-based RNA affinity tags offer both high specificity and the possibility to elute the mRNP complexes under native conditions. One of the most widely used RNA affinity purification methods is based on a stem-loop structure from the MS2 bacteriophage, which is specifically bound by the MS2 coat protein [2, 14]. In this method, several copies of the MS2 hairpin (<20 nucleotides) are fused to an RNA substrate and the RNA is incubated with a fusion protein between the MS2 coat protein and the maltose binding protein (MBP). The resultant RNP is then incubated with cell lysates/extract under proper conditions to assemble the mRNA processing complex. Subsequently, the assembled mRNP complexes are retrieved by using amylose beads, which specifically bind to the MBP moiety of the MBP-MS2 fusion protein. Finally, the bead-bound complexes can be eluted under native conditions and used for functional and/or structural studies. When used in combination with other purification steps, such as glycerol gradient centrifugation or gel filtration, this MS2-based RNA affinity method has enabled the purification of various spliceosomes (E, A, B, C and post-splicing mRNPs) [14–16] and mRNA 3′ processing complexes [2]. Therefore, RNA affinity purification has greatly facilitated the identification of novel mRNA processing factors and the study of the structure-function relationships of mRNA processing complexes. Next, we will provide a detailed protocol for applying this method for the purification of mRNA 3′ processing complex. It is important to point out that the protein- and RNA-affinity-based purification methods are not mutually exclusive and can be used in combination to obtain highly purified mRNP complexes.

2 Materials

2.1 RNA Substrate Preparation by In Vitro Transcription

1. DNA templates for 3MS2-tagged RNA substrates
   A plasmid that contains the coding sequence for your RNA substrate driven by a T7 or SP6 promoter. Three copies of the MS2 hairpin, the sequence of which is shown below, can be inserted either upstream or downstream of the RNA substrate sequence:

   5′-CGTACACCATCAGGGTACGAGCAAGCCCATTGCGTACACCATCAGGGTACGACTAGTACATTCGTACACCATCAGGGTACG-3′ (MS2 sequences are underlined)

   SV40 late poly(A) site (SVL) is a commonly used substrate for in vitro 3′ processing assays [2]. A mutant version of SVL with one nucleotide mutation in the poly(A) signal (AAUAAA to AACAAA) is deficient in recruiting mRNA 3′ processing factors and thus is used as a negative control [2]. The 3× MS2 affinity tag was inserted upstream of SVL or SVLm (Fig. 1). For additional considerations for designing an RNA substrate, *see* **Note 1**.
2. T7 or SP6 RNA polymerase.
3. Nucleotide triphosphate mix (NTPs) for making non-radiolabeled RNAs: 5 mM/each.

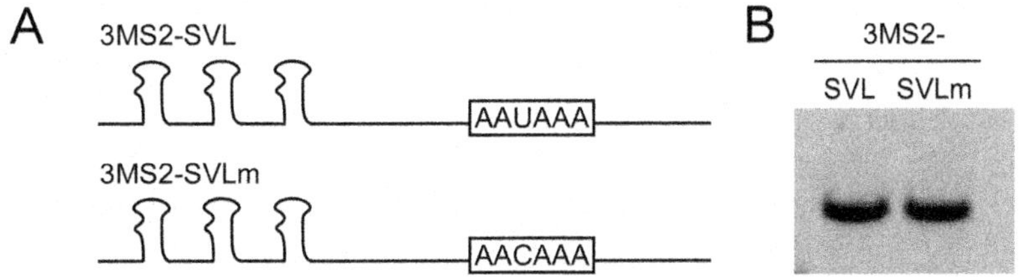

Fig. 1 (**a**) Schematic of poly(A) site and AAUAAA mutant tagged by 3× MS2 stem loops. (**b**) RNA detection of 3MS2-poly A site by 8% Urea-PAGE

4. NTPs for [α-^{32}P] UTP labeling: 1 mM UTP and 5 mM/each for ATP, CTP, and GTP.
5. [α-^{32}P] UTP, 10 μCi/μl.
6. 5× transcription buffer: 200 mM Tris (pH 7.9), 30 mM $MgCl_2$, 10 mM spermidine, and 50 mM NaCl.
7. 40 U/μl RNaseOUT or RNase inhibitor.
8. RNase-free DNase I.
9. Phenol/chloroform with isoamyl alcohol (PCI) (pH 4.3).
10. 100% ethanol and 75% ethanol.
11. GlycoBlue or glycogen (for RNA precipitation).
12. 10 M ammonium acetate.
13. 2× RNA loading buffer: 80% (v/v) formamide, 1× TBE, 0.1% (w/v) bromophenol blue, and 0.1% (w/v) xylene cyanol FF.
14. RNA elution buffer: 0.5 M ammonium acetate, 10 mM $MgCl_2$, 1 mM EDTA, and 0.2% SDS.
15. A spectrophotometer.
16. A liquid scintillation counter.
17. A phosphorimager.
18. Mineralight lamp (multiband UV-254/365 nm).

2.2 RNA Affinity Purification

1. pMBP-MS2 plasmid (Addgene: 65,104).
2. BL21 (DE3) *E. coli* strain.
3. Luria-Bertani (LB) liquid medium: 10 g NaCl, 10 g Tryptone, 5 g yeast extract and 950 ml H_2O, and autoclave it.
4. Lysis buffer for MS2-MBP purification: 20 mM Tris–HCl (pH 7.5), 200 mM NaCl, and 0.5 mM PMSF.
5. 1 M creatine phosphate (CP).
6. 10 μg/μl tRNA.
7. 40 U/μl RNaseOUT or RNase inhibitor.
8. HeLa nuclear extract (NE) [17].
9. 50 μg/μl heparin.
10. 10× Tris-Glycine: 500 mM Tris-base and 500 mM Glycine.
11. 4% PAGE: 1 ml of 40% Polyacrylamide gel solution (polyacrylamide: bisacrylamide = 80:1), 1 ml of 10× Tris-Glycin, 100 μl of 10% ammonium persulfate (APS), 10 μl TEMED and 7.9 ml H_2O.
12. TBE buffer: 90 mM Tris–HCl (pH 8.0), 90 mM borate, and 2 mM EDTA.
13. 5× Native gel loading buffer: 25% (v/v) Glycerol in TBE buffer with 0.25% (w/v) bromophenol blue.

14. 10% glycerol gradient solutions: 20 mM HEPES (pH 7.9), 60 mM KCl, 1 mM $MgCl_2$, 0.1 mM EDTA, 10% glycerol, and fresh 0.5 mM DTT.
15. 30% glycerol gradient solutions: 20 mM HEPES (pH 7.9), 60 mM KCl, 1 mM $MgCl_2$, 0.1 mM EDTA, 30% glycerol, and fresh 0.5 mM DTT.
16. RNA pulldown buffer: 20 mM Hepes-KOH, pH 7.9, 100 mM KCl, 1 mM $MgCl_2$, 1% Triton X-100, and fresh 0.5 mM DTT.
17. Maltose.
18. Amylose resin.
19. Prechilled acetone.

3 Methods

3.1 Prepare RNA Substrates by In Vitro Transcription

1. Linearize the template DNA plasmid by digesting 1 μg of the plasmid with a restriction enzyme that cuts immediately downstream of the RNA substrate sequence.
2. Purify the digested DNA and dissolve in 10 μl H_2O.
3. Set up in vitro transcription reaction as follows:

   10 μl of DNA template (1 μg).

   2.5 μl of 100 mM DTT.

   2.5 μl of NTP mix (5 mM of each NTP).

   0.5 μl of 5 U/μl RNaseOUT.

   5 μl of 5× Transcription buffer.

   1 μl of T7/SP6 polymerase.

   3.5 μl of H2O.

   25 μl (total volume).

   The above reaction is for making unlabeled RNAs. For making radiolabeled RNA: set up in vitro transcription as described above with the following modifications: add 1–3 μl [α-^{32}P] UTP and use an NTP mix that contains 1 mM of UTP and 5 mM/each of ATP, GTP, and CTP (*see* **Notes 2** and **3**).
4. Incubate at 37 °C for 2 h.
5. Add 1 μl of RNase-free DNase I and incubate at 37 °C for another 15 min.
6. Add 175 μl of H_2O and 200 μl PCI and vortex for 10 s.
7. Spin down at 16,000 × *g* for 15 min at 4 °C.
8. Retrieve 180 μl supernatant, and add 50 μl 10 M ammonium acetate, 750 μl of 100% ethanol, 1 μl of glycogen, and mix well.
9. Put on dry ice for at least 30 min.
10. Spin at 16,000 × *g* for 20 min at 4 °C. A pellet should be clearly visible at the bottom of the tube.

11. Remove the supernatant and wash the pellet with cold 75% ethanol.
12. Spin at 12,000 × *g* for 10 min at 4 °C.
13. Air dry the pellet for 10 min at room temperature.
14. Resuspend RNA pellet in 50 μl H_2O.
15. Measure RNA concentration using a NanoDrop spectrophotometer. For the quantification of radiolabeled RNA, do not use a spectrophotometer to measure concentration! Instead, take out 1% of the in vitro transcription reaction (R_{before}) and 1% of the final RNA (R_{after}). Measure the radiation levels of R_{before} and R_{after} with a liquid scintillation counter and calculate the molar amount of the radiolabeled RNA product using the following formula:
    RNA amount = (2.5 μl × 1 mM × R_{after}/R_{before})/(the number of Us in the RNA)
16. Resolve 200 ng RNA on 8% Urea-polyacrylamide gel in 1× RNA loading buffer.
17. Stain the gel with SYBR green for 30 min in the dark to check the quality of RNA.
18. (***Optional step***) If a significant amount of degraded or truncated RNAs are detected, gel purify the RNA to obtain the full-length fragment. To this end, resolve the RNA products from the in vitro transcription reaction on an 8% urea acrylamide gel, visualize the RNA band(s) with a phosphorimager (for radiolabeled RNA) or long wave light of ultraviolet light (for unlabeled RNA), and cut out the band containing the full-length RNA. Put the gel piece in a 1.5 ml tube with 400 μl RNA elution buffer. Rotate the tube at 37 °C for 1 h. and PCI extract twice. Precipitate the RNA as described above.

3.2 Preparation of MBP-MS2 Fusion Protein

1. Transform pMBP-MS2 into BL21 (DE3) strain.
2. Pick a single colony and inoculate 2 ml LB plus 100 μg/ml ampicillin.
3. Incubate the starter culture in a shaker (200 rpm) at 37 °C overnight.
4. Transfer 500 μl of the starter culture into 500 ml LB plus 100 μg/ml ampicillin.
5. Incubate in a shaker (200 rpm) at 37 °C until $O.D_{600}$ reaches ~0.5.
6. Add IPTG to a final concentration of 1 mM.
7. Further incubate in a shaker at 37 °C for 3 h.
8. Harvest cells by centrifugation at 5000 × *g* for 10 min.
9. Carefully decant the media and resuspend the cell pellet in 40 ml prechilled PBS buffer, and transfer to a 50 ml falcon tube.

10. Spin at 5000 × *g* for 10 min.
11. Carefully decant the supernatant and resuspend the cell pellet in 10 ml lysis buffer.
12. Sonicate the cells in 5 ml aliquots (output 3, 10 s for four times) on ice.
13. Spin at 9400 × *g* for 20 min.
14. (***Optional step***) As MBP-MS2 protein tends to bind to nucleic acids, which may interfere with the purification and subsequent analyses, it is recommended to remove the nucleic acids by adding $CaCl_2$ to 10 mM and 5 μl of micrococcal nuclease to the cleared lysate.
15. Wash 500 μl amylose beads with lysis buffer and then mix the beads with the cell lysate.
16. Rotate for 1 h at 4 °C.
17. Load the mix onto a column and let the cell lysate pass through the column by gravity flow.
18. Wash the beads with 5 ml lysis buffer for three times.
19. Wash the column with 10 ml of PBS.
20. Elute the MBP-MS2 protein by adding 500 μl elution buffer (PBS + 15 mM maltose) to the beads. Incubate for 2 min, spin at 1000 × *g* for 5 min, and collect the supernatant. Repeat this step until no protein can be detected in the eluate by using Bradford assay. Usually, all proteins are eluted in the first 3 elutions.
21. Dialysis against 2 l of Buffer D twice for at least 2 h each time at 4 °C.
22. Measure protein concentration using Bradford assay and store purified protein in small aliquots at −80 °C.

3.3 Test the Interaction Between MBP-MS2 and RNA Substrate In Vitro

It is necessary to check the binding ability and optimize the ratio of MS2-MBP protein to RNA. Excess RNA or protein maybe reduces pulldown efficiency to some extent.

1. Mix 2 pmol RNA substrate with 10 ~ 100 pmol MS2-MBP (use several different amounts) in 10 μl RNA pulldown buffer, respectively.
2. Incubate on ice for 20 min.
3. Add 2.5 μl 5× native gel loading buffer and mix well.
4. Resolve 10 μl on a 1.5% agarose gel and identify the lowest amount of protein that can shift all of the RNAs. (For radioactivity labeled RNA, 4% PAGE (80:1) is a better choice and check signal by phosphorimaging. Run the gel in 1× tris-glycine buffer in a cold room at 100 v for 210 min, and then expose to phosphorimager.)

   For 3MS2-SVL and 3MS2-SVLm, the molar ratio of 1:20 between RNA and MS2-MBP is optimal.

3.4 RNA Affinity Purification of mRNA 3′ Processing Complex

1. Set up the cleavage/polyadenylation reaction as follows: 5 pmol of 3MS2-SVL or 3MS2-SVLm RNA, 100 pmol of MS2-MBP, and add RNA pulldown buffer to bring the total volume up to 50 μl.
2. Incubate for 30 min on ice.
3. Add 5 μl of 0.1 M ATP, 10 μl of 1 M CP, 5 μl of 10 μg/μl tRNA, 200 μl NE, 230 μl H_2O and mix well (*see* **Note 4**).
4. Incubate for 20 min at 30 °C.
5. Put sample on ice and add 4 μl of 50 μg/μl heparin to mix well (*see* **Note 5**).
6. Add 50 μl prewashed amylose beads to the reaction and rotate for 1 h at 4 °C (*see* **Note 6**).
7. Spin the mixture at 2000 × *g* for 2 min and remove the supernatant.
8. Add 1 ml RNA pulldown buffer and rotate for 15 min.
9. Spin the mixture at 2000 × *g* for 5 min and remove the supernatant.
10. Repeat the wash three times.
11. Elute complex by adding 100 μl elution buffer (RNA pull-down buffer plus 12 mM maltose) to the beads and rotate for 20 min at 4 °C.
12. Spin down for 2 min at 2000 × *g* and transfer the supernatant to a new 1.5 ml EP tube.
13. Repeat **step 11** and combine eluates.
14. Add 800 μl cold acetone and keep it at −20 °C for 20 min (up to overnight).
15. Spin down proteins at 16,000 × *g* for 20 min at 4 °C.
16. Air dry pellet for 10 min at room temperature.
17. The protein pellet can be analyzed by mass spectrometry to identify mRNA processing factors, or by SDS-PAGE and silver staining (Fig. 2) or Western blotting analyses.

3.5 Glycerol Gradient Sedimentation (Optional Step)

When a substrate RNA is incubated with cell extracts, multiple distinct complexes may form, including heterogeneous complexes that consist of mostly hnRNPs, and mRNA processing complexes of different stages (ref for both spliceosome and 3′ processing complexes). Although the one-step affinity purification method is usually sufficient for identification of mRNA processing factors (provided that proper negative control samples are used), additional purification steps may be necessary to obtain a homogenous population of mRNA processing complexes for structural and/or functional studies. Gradient sedimentation resolves particles according to their sizes and is suitable for further

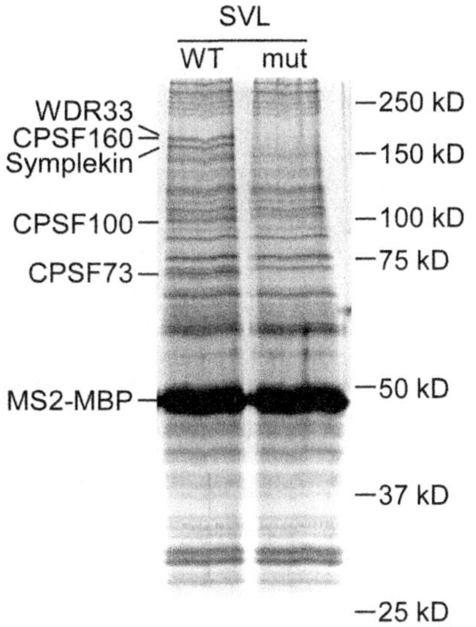

Fig. 2 Poly(A) factors pull downed by 3MS2-SVL poly(A) site RNA. Silver staining was performed after RNA pulldown. Only wild-type poly(A) site can pull down poly(A) factors, which are indicated by their respective names

purification of mRNA processing complexes. Detailed protocols are described below:

1. Make a 10–30% glycerol gradient (12 ml) using a gradient maker in SW41 centrifuge tubes, and leave it at 4 °C for 1 h or more.
2. Carefully load the in vitro mRNA processing reaction on the top of the gradient.
3. Load the gradient onto an ultracentrifuge and spin at 80,000 × *g* for 15 h.
4. Manually collect 0.5 ml fractions from the top.
5. Measure radioactivity of each fraction.
6. Pick specific peak fractions that contain your desired mRNA processing complexes (combine fractions if multiple fractions are found in a peak) and mix with 50 μl amylose beads that have been prewashed with wash buffer.
7. Rotate for 30 min (up to 2 h) at 4 °C.
8. Follow **step 7** of Subheading 3.4.

4 Notes

1. Several factors need to be considered for designing RNA substrates. First, the RNA substrate sequence has to be tested beforehand to ensure that it allows the assembly of the desired processing complex in vitro. Second, the 3MS2-tagged RNA substrates should be carefully compared with the untagged RNA to ensure that the tag does not interfere with the processing reaction.

Third, as many cellular proteins bind promiscuously to RNAs, it is essential to include negative control RNA substrates for affinity purification. The ideal negative controls are mutant versions of the RNA substrates, which contain a small number of substitutions that abolish the assembly of the processing complex.

2. For glycerol gradient sedimentation, it is necessary to use [α-^{32}P] UTP-labeled RNA to determine in which fractions the 3′ processing complex are by measuring the intensity of radioactivity. When the formal assay is conducted non-radioactivity labeled RNA are used, and it is easy to pick certain fractions the target complex is in.
3. Lots of RNA are required for this assay and T7 RNA polymerase can meet this demand. For poly(A) RNA substrate used in this chapter, about 7 μg RNA can be obtained from a 25 μl transcription reaction. Scale up if more RNA is required for the further assay.
4. Pulldown efficiency depends on interaction ability between RNA and relevant complexes. The amount of proteins from 500 μl RNA pulldown reaction in this chapter can be used for ten western blot analyses. Reaction volume can be scaled up based on further assay.
5. Addition of heparin is critical for reducing background. However, excessive heparin also decreases pulldown efficiency. The final concentration depends on the motif strength in RNA. For example, more heparin can be used for stronger poly(A) site. It is necessary to optimize heparin concentration for each poly(A) site.
6. Poly-prep chromatography columns are recommended if the reaction is more than 2 ml. It is very convenient to use column after incubation at 37 °C for incubation with amylose resin, washing, and elution.

Acknowledgment

This study was supported by grants from the National Institutes of Health (GM090056 and CA177651) and the American Cancer Society (RSG-12-186).

References

1. Will CL, Luhrmann R (2011) Spliceosome structure and function. Cold Spring Harb Perspect Biol 3(7):a003707
2. Shi Y, Di Giammartino DC, Taylor D, Sarkeshik A, Rice WJ, Yates JR 3rd, Frank J, Manley JL (2009) Molecular architecture of the human pre-mRNA 3′ processing complex. Mol Cell 33(3):365–376
3. Matera AG, Wang Z (2014) A day in the life of the spliceosome. Nat Rev Mol Cell Biol 15(2):108–121
4. Grote M, Wolf E, Will CL, Lemm I, Agafonov DE, Schomburg A, Fischle W, Urlaub H, Luhrmann R (2010) Molecular architecture of the human Prp19/CDC5L complex. Mol Cell Biol 30(9):2105–2119

5. Khanna M, Van Bakel H, Tang X, Calarco JA, Babak T, Guo G, Emili A, Greenblatt JF, Hughes TR, Krogan NJ, Blencowe BJ (2009) A systematic characterization of Cwc21, the yeast ortholog of the human spliceosomal protein SRm300. RNA 15(12):2174–2185
6. Tardiff DF, Rosbash M (2006) Arrested yeast splicing complexes indicate stepwise snRNP recruitment during in vivo spliceosome assembly. RNA 12(6):968–979
7. Yan C, Hang J, Wan R, Huang M, Wong CC, Shi Y (2015) Structure of a yeast spliceosome at 3.6-angstrom resolution. Science 349(6253): 1182–1191
8. Leppek K, Stoecklin G (2014) An optimized streptavidin-binding RNA aptamer for purification of ribonucleoprotein complexes identifies novel ARE-binding proteins. Nucleic Acids Res 42(2):e13
9. Iioka H, Loiselle D, Haystead TA, Macara IG (2011) Efficient detection of RNA-protein interactions using tethered RNAs. Nucleic Acids Res 39(8):e53
10. Hartmuth K, Urlaub H, Vornlocher HP, Will CL, Gentzel M, Wilm M, Luhrmann R (2002) Protein composition of human prespliceosomes isolated by a tobramycin affinity-selection method. Proc Natl Acad Sci U S A 99(26):16719–16724
11. Keryer-Bibens C, Barreau C, Osborne HB (2008) Tethering of proteins to RNAs by bacteriophage proteins. Biol Cell 100(2):125–138
12. Rodgers JT, Patel P, Hennes JL, Bolognia SL, Mascotti DP (2000) Use of biotin-labeled nucleic acids for protein purification and agarose-based chemiluminescent electromobility shift assays. Anal Biochem 277(2):254–259
13. Hou S, Shi L, Lei H (2016) Biotin-streptavidin affinity purification of RNA-protein complexes assembled in vitro. Methods Mol Biol 1421: 23–34
14. Zhou Z, Sim J, Griffith J, Reed R (2002) Purification and electron microscopic visualization of functional human spliceosomes. Proc Natl Acad Sci U S A 99(19):12203–12207
15. Bessonov S, Anokhina M, Will CL, Urlaub H, Luhrmann R (2008) Isolation of an active step I spliceosome and composition of its RNP core. Nature 452(7189):846–850
16. Yoshimoto R, Kataoka N, Okawa K, Ohno M (2009) Isolation and characterization of post-splicing lariat-intron complexes. Nucleic Acids Res 37(3):891–902
17. Dignam JD, Lebovitz RM, Roeder RG (1983) Accurate transcription initiation by RNA polymerase II in a soluble extract from isolated mammalian nuclei. Nucleic Acids Res 11(5):1475–1489

Chapter 6

Polyadenylation Site-Based Analysis of Transcript Expression by 3′READS+

Dinghai Zheng and Bin Tian

Abstract

Deep sequencing of the 3′ end region of poly(A)+ RNA identifies the cleavage and polyadenylation site (PAS) and measures transcript abundance. However, mispriming at internal A-rich regions by the oligo-dT oligo in reverse transcription can lead to falsely identified PASs. This problem can be resolved by direct ligation of an adapter to the 3′ end of RNA. However, ligation-based methods are often inefficient. Here, we describe 3′READS+, an accurate and sensitive method for deep sequencing of the 3′ end of poly(A)+ RNA. Through partial digestion by RNase H of the poly(A) tail bound to a locked nucleic acid (LNA)/DNA hybrid oligo, this method sequences an optimal number of terminal A's, which balances sequencing quality and accurate identification of PAS in A-rich regions. With efficient ligation steps, 3′READS+ is amenable to small amounts of input RNA. 3′READS+ can also be readily used as a cost-effective method for gene expression analysis.

Key words Alternative cleavage and polyadenylation, Deep sequencing, RNA-seq, 3′ end sequencing, 3′READS+

1 Introduction

Almost all eukaryotic mRNAs have a poly(A) tail, which is important for their metabolism in the cell. Cleavage and polyadenylation (C/P) is the mechanism responsible for adding the poly(A) tail and is coupled with termination of transcription [1]. C/P is also responsible for 3′ end maturation of most, if not all, long noncoding RNAs that are transcribed by RNA polymerase II. Most eukaryotic genes harbor multiple C/P sites, or polyadenylation sites (PASs), leading to expression of alternative polyadenylation (APA) isoforms [2, 3].

Deep sequencing of RNA (RNA-seq) provides quantitative measurement of RNA abundance with single-nucleotide resolution [4]. Sequencing at the 3′ end region of poly(A)+ RNA identifies the PAS and hence is of use for the analysis of APA. In addition, 3′ end sequencing can be used as a cost-effective way for gene

Yongsheng Shi (ed.), *mRNA Processing: Methods and Protocols*, Methods in Molecular Biology, vol. 1648,
DOI 10.1007/978-1-4939-7204-3_6, © Springer Science+Business Media LLC 2017

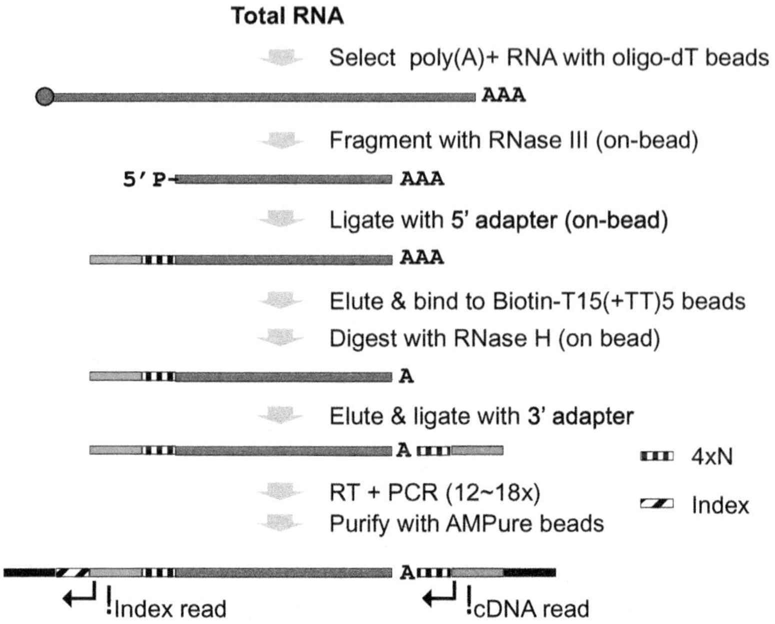

Fig. 1 Work flow of the 3′READS+ protocol. AAA, poly(A) tail; A, shortened poly(A) tail (to ~13 nt); *green box*, RNA; *blue box*, 5′ adapter; *orange box*, 3′ adapter; Four random nucleotides in the adapters and index (*barcode*) sequence from PCR primers are also indicated

expression analysis, as first demonstrated by the Significance Analysis of Gene Expression (SAGE) technique [5].

A number of methods for 3′ end sequencing have been developed in past few years, which catapulted our understanding of APA genome-wide [6–10]. Methods using oligo(dT) can lead to falsely identified PASs due to mispriming at internal A-rich sequences [11]. While bioinformatic methods can be used to partially address this [12], they can also discard bona fide PASs. Ligation-based methods solve the mispriming problem because reverse transcription of RNA is initiated from the 3′ adapter added by ligation. However, inefficient ligation hampers the use of these methods when a limited amount of RNA is available [9, 10].

To increase sensitivity and accuracy of PAS identification, we developed 3′READS+ [13], the overall protocol of which is illustrated in Fig. 1. The method uses an LNA/DNA hybrid oligo to protect ~13 As at the beginning of the poly(A) tail in an RNase H digestion reaction, allowing accurate identification of genuine PASs, even when situated in A-rich genomic regions, and avoiding a longer A-sequence that could compromise sequencing quality on current Illumina sequencing platforms. By separating 5′ and 3′ adapter ligation steps, we substantially improved the ligation efficiency. The current version of 3′READS+ can comfortably handle

100 ng of total RNA as input. Additional features to address sequence bias caused by RNA ligation and over amplification of cDNA by PCR are also incorporated into the design of adapter sequences.

2 Materials

Prepare all solutions using Diethylpyrocarbonate (DEPC)-treated ultrapure water (ddH_2O), and analytical grade reagents.

2.1 Poly(A)+ RNA Isolation

1. Oligo $d(T)_{25}$ coated magnetic beads.
2. Magnetic stand.
3. Micro-centrifuge.
4. 70 °C water bath.
5. 2× Binding buffer: 20 mM Tris–HCl, pH 7.5, 300 mM NaCl, 2 mM EDTA, 0.1% TWEEN 20. Mix 1 ml of 1 M Tris–HCl, pH 7.5, 6 ml of 2.5 M NaCl, 0.2 ml of 0.5 M EDTA, pH 8.0, and 50 μl of TWEEN 20. Make up to 50 ml with water. Store at 4 °C.

2.2 RNA Fragmentation

1. RNase III (1 U/μl).
2. RNase III buffer: 10 mM Tris–HCl, pH 8.3, 60 mM NaCl, 10 mM $MgCl_2$, 1 mM DTT. Mix 0.5 ml of 1 M Tris–HCl, pH 8.3, 1.2 ml of 2.5 M NaCl, and 0.5 ml of 1 M $MgCl_2$. Make up to 49.95 ml with water. Store at 4 °C. Before use, take out the amount needed and add 1/1000 volume of 1 M DTT.
3. Magnetic stand.
4. 37 °C water bath.

2.3 5′ Adapter Ligation

1. T4 RNA Ligase 1 (10 U/μl), 50% PEG 8000, and T4 RNA ligase buffer.
2. 10 mM ATP.
3. RNase inhibitor (20 U/μl).
4. 5′ adapter: 5 μM RNA oligo with the sequence 5′-CCUUGG CACCCGAGAAUUCCANNNN. Dissolved in water, aliquoted, and saved at −80 °C.
5. Rotator at room temperature.

2.4 Partial Poly(A) Tail Digestion

1. Streptavidin-coated magnetic beads. (10 μg/μl).
2. Magnetic stand.
3. RNase H (10 U/μl).
4. DNA/LNA oligo: biotin-TTTTTTTTTTTTTTT+TT+TT+TT+TT+TT, where "+T" denotes LNA. Dissolve in ddH2O and make a 20 μM solution. Store at −20 °C.

5. Bead prewash buffer: 100 mM NaOH, 50 mM NaCl and 0.05% Tween 20. Mix 0.5 ml of 10 M NaOH, 1 ml of 2.5 M NaCl. Make up to 50 ml with ddH_2O. Store at 4 °C. Before use, add 1/2000 volume of TWEEN 20 (final concentration 0.05% v/v).
6. Coupling buffer: 10 mM 1 M Tris–HCl, pH 7.5, 1 M NaCl, 1 mM EDTA, and 0.05% TWEEN 20. Mix 0.5 ml of 1 M Tris–HCl, pH 7.5, 20 ml of 2.5 M NaCl, 100 μl of 0.5 M EDTA, pH 8.0 and 25 μl of TWEEN 20. Make up to 50 ml with ddH_2O. Store at 4 °C.
7. pH 7.5 TE buffer: 10 mM Tris–HCl, pH 7.5, and 1 mM EDTA. Mix 0.5 ml of 1 M Tris–HCl, pH 7.5 and 100 μl of 0.5 M EDTA. Make up to 50 ml with ddH_2O. Store at 4 °C.
8. Wash buffer: 10 mM Tris–HCl, pH 7.5, 1 mM NaCl, 1 mM EDTA, and 0.05% TWEEN 20. Mix 0.5 ml of 1 M Tris–HCl, pH 7.5, 20 μl of 2.5 M NaCl, 100 μl of 0.5 M EDTA, pH 8.0, 25 μl of TWEEN 20. Make up to 50 ml with ddH_2O. Store at 4 °C.
9. RNase H buffer: 50 mM Tris–HCl, pH 7.5, 5 mM NaCl, 10 mM $MgCl_2$, and 10 mM DTT. Mix 2.5 ml of 1 M Tris–HCl, pH 7.5, 100 μl of 2.5 M NaCl, 500 μl of 1 M $MgCl_2$. Make up to 49.5 ml with ddH_2O. Store at 4 °C. Before use, add 1/100 volume of 1 M DTT.
10. Elution buffer: 1 mM NaCl, 1 mM EDTA, and 0.05% TWEEN 20. Mix 20 μl of 2.5 M NaCl, 100 μl of 0.5 M EDTA, pH 8.0, 25 μl of TWEEN 20. Make up to 50 ml with ddH_2O. Store at 4 °C.
11. 50 °C water bath.
12. 1 μg/μl Glycogen.
13. 3 M Sodium Acetate, pH 5.2.
14. 200 proof ethanol.
15. 75% ethanol. Mix 37.5 ml 200 proof ethanol and 12.5 ml water. Store at −20 °C.
16. Refrigerated micro-centrifuge.

2.5 3′ Adapter Ligation

1. T4 RNA Ligase 2, truncated KQ (200 U/μl).
2. 50% PEG 8000 solution.
3. T4 RNA ligase buffer.
4. RNase inhibitor (20 U/μl).
5. 5′ adenylated 3′ blocked 3′ adapter with degenerated nucleotides: 5′-rApp/NNNNGATCGTCGGACTGTAGAACTCTGAAC/3ddC (10 μM).
6. 25 °C water bath.

2.6 Reverse Transcription

1. M-MLV reverse transcriptase (200 U/μl) and buffer.
2. dNTP (25 mM each).
3. Reverse transcription primer: 5′-GTTCAGAGTTCTACAG TCCGACGATC (5 μM).
4. Thermal cycler.

2.7 PCR

1. DNA Polymerase (2 U/μl) and buffer.
2. dNTP (25 mM each).
3. Reverse PCR primer (GX3): 5′-AATGATACGGCGACCA CCGAGATCTACACGTTCAGAGTTCTACAGTCCGA (25 μM).
4. Indexed forward PCR primers (index region in bracket): 5′-CAAGCAGAAGA CGGCATACGAGAT[NNNNNN]GT GACTGGAGTT CCTTGGCACCCGAGAATTCCA (5 μM).
5. Thermal cycler.

2.8 Size Selection

1. Solid phase reversible immobilization (SPRI) beads.
2. Magnetic stand.
3. Freshly prepared 80% ethanol. Mix 8 ml 200 proof ethanol with 2 ml water.
4. Elution buffer: 10 mM Tris–HCl, pH 8.0. Mix 10 μl of 1 M Tris–HCl, pH 8.0 with 90 μl of ddH_2O.

2.9 Library Quantification and Normalization

1. Agilent Bioanalyzer or TapeStation.
2. High Sensitivity DNA analysis kit or High Sensitivity D1000 kit.
3. Qubit and dsDNA HS Assay Kit.

3 Methods

3.1 Poly(A)+ RNA Isolation

1. Transfer resuspended oligo(dT)$_{25}$ magnetic bead slurry to a 1.5 ml tube on a magnetic stand (*see* **Note 1**). Use 10 μl for each sample. Wait for the liquid to become clear and remove the supernatant.
2. Wash the beads with 500 μl of 2× Binding Buffer twice.
3. Resuspend the beads in 2× Binding Buffer, with 50 μl per sample.
4. Dilute 0.1 ~ 15 μg of high-quality total RNA to 50 μl with RNase-free ddH_2O in a 1.5 ml tube (*see* **Note 2**). Incubate RNA samples in a 70 °C water batch for 2 min and then immediately transfer the samples onto ice.

5. Add 50 μl of the beads prepared in **step 3** to each sample, and agitate the tubes to mix. Incubate the samples at room temperature on a rotator for 30 min.
6. After a quick spin, put the tubes on the magnetic stand and discard the supernatant. Wash the beads in each tube with 500 μl of 1× Binding Buffer three times.

3.2 On-Bead RNA Fragmentation

1. Prepare RNase III Digestion Master Mix by mixing 14.25 μl of RNase III Buffer with 0.75 μl of (1 U/μl) of RNase III per sample. Keep the Master Mix on ice.
2. Wash the beads from Subheading 3.1 with 500 μl RNase III Buffer once. After a quick spin, put the tubes back onto the magnetic stand, and remove residual buffer using a 200 μl pipette.
3. Aliquot 15 μl of the RNase III Master Mix to each tube and completely resuspend the beads by pipetting. Keep the tubes on ice during this process.
4. Transfer the tubes to a 37 °C water bath. Incubate for 15 min, with gentle agitation once in the middle of the incubation to prevent bead precipitation.
5. Transfer the tubes onto ice. Immediately add 0.5 ml of 1× Binding Buffer to each tube, agitate the tubes to mix, and put the tubes back to the magnetic stand. Discard the supernatant. Wash the beads with 0.5 ml of 1× Binding two more times.
6. After a quick spin, put the tubes back on the magnetic stand, and remove residual buffer using a 200 μl pipette.

3.3 5′ Adapter Ligation (see Note 4)

1. Prepare 5′ adapter ligation master mix. First, incubate 5′ adapter (5 μM, 2 μl per sample) in a 70 °C water bath for 2 min and immediately put the tube on ice. Next, add to the tube the following (per sample): 2.5 μl of ddH_2O, 1.5 μl of 10 mM ATP, 1.5 μl of 10× T4 RNA ligase buffer, 6 μl of 50% PEG 8000, 0.5 μl of RNase inhibitor, and 1 μl of RNA ligase 1. Pipette to mix well.
2. Aliquot 15 μl of the master mix to each tube from Subheading 3.2 and completely resuspend the beads by pipetting.
3. Keep the tubes on a room temperature rotator and incubate for 2 h.

3.4 Conjugation of the DNA/LNA Oligo to Streptavidin-Coated Magnetic Beads

1. Transfer completely resuspended streptavidin-coated magnetic beads slurry to a 1.5 ml tube on a magnetic stand, 12 μl per sample. Wait for the liquid to become clear and remove the supernatant.
2. Wash the beads with 500 μl of Bead Prewash Buffer twice.

3. Wash the beads once with 500 μl of Coupling Buffer. Resuspend the beads in 100 μl per sample of Coupling Buffer.
4. Add 0.5 μl per sample of 20 μM DNA/LNA oligo to the resuspended beads and immediately invert the tube several times. Incubate at room temperature on a rotator for 30 min.
5. Capture the beads on a magnetic stand and discard the supernatant. Wash the beads three times with 0.5 ml of 1× Binding Buffer. Resuspend the beads in 100 μl per sample of 2× Binding Buffer.

3.5 Binding of the Ligation Products to the DNA/LNA Oligo on Magnetic Beads

1. Add 85 μl of pH 7.5 TE to the ligation reaction tubes from Subheading 3.3. Mix well by pipetting.
2. Incubate the tubes at 70 °C for 2 min to completely elute RNA from the beads and immediately transfer the tubes to the magnetic stand. Transfer the supernatant to new tubes on ice.
3. Aliquot 100 μl of the streptavidin-coated magnetic beads prepared in **step 5** of Subheading 3.4 to each tube of eluted RNA and mix by agitation. Incubate at room temperature on a rotator for 30 min.
4. Transfer the tubes onto the magnetic stand and discard the supernatant. Wash the beads with 0.5 ml of Wash Buffer three times.

3.6 On-Bead Partial Poly(A) Tail Digestion

1. Prepare RNase H reaction by mixing 30 μl of RNase H Buffer and 0.03 μl (0.3 U) of RNase H for each sample.
2. Wash the beads from Subheading 3.4 with 0.5 ml of RNase H Buffer once. After a quick spin, put the tubes back on the magnetic stand, and remove residual buffer using a 200 μl pipette.
3. Aliquot 30 μl of the RNase H reaction to each tube and completely resuspend the beads by pipetting. Keep the tubes on ice during this process.
4. Transfer the tubes to a 37 °C water bath. Incubate for 30 min with gentle agitation every 10 min to prevent bead precipitation.
5. Transfer the tubes onto ice. Immediately add 500 μl of 1× Wash Buffer to each tube to stop enzyme reaction, mix by agitating the tubes, and transfer the tubes to the magnetic stand. Discard the supernatant and wash the beads with 500 μl of Wash Buffer two more times.
6. After a quick spin, put the tubes back on the magnetic stand, and remove residual buffer using a 200 μl pipette.
7. Resuspend beads in each tube with 50 μl of Elution Buffer by pipetting. Keep the tubes on ice during this process.
8. Incubate in a 50 °C water bath for 2 min, and immediately transfer the sample to a magnetic stand. Collect the supernatant in 1.5 ml tubes on ice.

9. Repeat the elution once and combine the eluted RNAs in the tubes on ice.
10. Centrifuge the eluted RNA at maximum speed at 4 °C for 1 min to collect carried-over beads. Put the tubes on a magnetic stand and transfer the supernatant to new 1.5 ml tubes.
11. Mix the eluted RNA in each tube with 10 μl 3 M NaAc, 1 μl glycogen and 300 μl of ethanol. Incubate the mixture at −20 °C for at least 2 h or at −80 °C for 45 min.
12. Centrifuge the tubes at 16,000 × *g* for 20 min at 4 °C. Discard the supernatant and wash the pellet with 500 μl of cold 75% ethanol. Centrifuge the samples at 16,000 × *g* again for 5 min at 4 °C. Discard the supernatant and air-dry the pellet at room temperature.
13. Dissolve the RNA fragments in 6 μl of ddH_2O.

3.7 3′ Adapter Ligation

1. Transfer 0.5 μl (per sample) of 5 μM 3′ adapter to a 1.5 ml tube. Incubate the adapter in a 70 °C water bath for 2 min and immediately put the tube on ice.
2. Make 3′ adapter ligation master mix. Add to the 3′ adapter tube the following per sample: 6 μl of 50% PEG 8000, 1.5 μl of 10× T4 RNA ligase buffer, 0.5 μl of RNase inhibitor, and 0.5 μl of RNA ligase 2 truncated, KQ. Mix reagents thoroughly by pipetting.
3. Incubate the RNA from **step 13** of Subheading 3.6 in a 70 °C water bath for 2 min and immediately put the tube on ice.
4. Aliquot 9 μl of the 3′ adapter ligation master mix with 6 μl of the RNA. Mix reagents thoroughly by pipetting.
5. Incubate the samples in a 25 °C water bath for 2 h.
6. Add 85 μl of ddH_2O, 10 μl 3 M NaAc, 1 μl glycogen, and 300 μl of ethanol to each tube. Incubate the mixture at −20 °C for at least 2 h or −80 °C for 45 min.
7. Centrifuge the tubes at 16,000 × *g* for 20 min at 4 °C. Discard the supernatant and wash the pellet with 500 μl of cold 75% ethanol. Centrifuge the samples at 16,000 × *g* again for 5 min at 4 °C. Discard the supernatant and air-dry the pellet at room temperature.

3.8 Reverse Transcription

1. Prepare RT Primer Mix, which contains 6.15 μl of ddH_2O, 0.25 μl of 25 mM dNTP, and 0.6 μl of 5 μM 3′ directional RT primer for each sample.
2. Dissolve RNA pellet from **step** 7 of Subheading 3.7 with 7 μl of the RT Primer Mix. Incubate the tubes at 70 °C for 2 min and then immediately transfer them on ice for 1 min.

3. Prepare RT Master Mix, which contains, for each reaction, 0.8 μl of ddH_2O, 2 μl of 5× M-MLV RT buffer, 0.1 μl of RNase inhibitor and 0.5 μl of M-MLV RT. Mix well by gentle vortex.
4. Add 3 μl of the master mix to tube on ice. Tap the tubes to mix and do a quick spin to collect the solution.
5. Carry out reverse transcription reaction in a 42 °C water bath for 60 min, followed by inactivation of the reverse transcriptase at 70 °C for 20 min.

3.9 PCR

1. Transfer 8 μl of the RT reaction into each PCR tube.
2. Add 5 μl of 5 μM barcoded reverse PCR primer to the corresponding PCR tube.
3. Prepare PCR Master Mix by mixing in a 1.5 ml tube the following for each sample: 25 μl of ddH_2O, 10 μl of 5× PCR Buffer, 1 μl of 25 μM GX3 primer, 0.5 μl of 25 mM dNTP Mix, and 0.5 μl of DNA Polymerase. Mix by gentle vortex, followed by a quick spin.
4. Aliquot 37 μl of the PCR Master Mix into each of the RT tube. Mix by gentle agitation, followed by a quick spin.
5. Run the PCR reaction with the following program: 98 °C for 30 s, 12 ~ 18 cycles of (98 °C for 10 s, 60 °C for 30 s and 72 °C for 15 s), 72 °C for 60 s and hold at 4 °C. The lower the input RNA amount, the higher PCR cycle number should be used. 15 cycles of PCR are usually enough for 1 μg of input total RNA.

3.10 Library Purification Using SPRI Beads

1. Mix 30 μl of SPRI beads with 50 μl of the PCR product in a 1.5 ml tube. Incubate the sample at room temperature for 5 min.
2. After a quick spin, place the tubes on a magnetic stand. After the solution becomes clear, transfer the supernatant from each tube to a new 1.5 ml tube. Discard the tubes containing used beads, which contain PCR products longer than 500 bp.
3. Add 20 μl of SPRI beads to the supernatant in each tube. Mix well by pipetting up and down a few times. Incubate at room temperature for 5 min. Place the tubes on a magnetic stand. After the solution becomes clear, discard the supernatant, which contains PCR products with short inserts.
4. Add 100 μl of freshly prepared 80% ethanol to the tubes containing magnetic beads while keeping the tubes on a magnetic stand. Incubate at room temperature for 30 s, and then remove and discard the supernatant.
5. Repeat the bead wash once and air-dry the beads for 30 min.

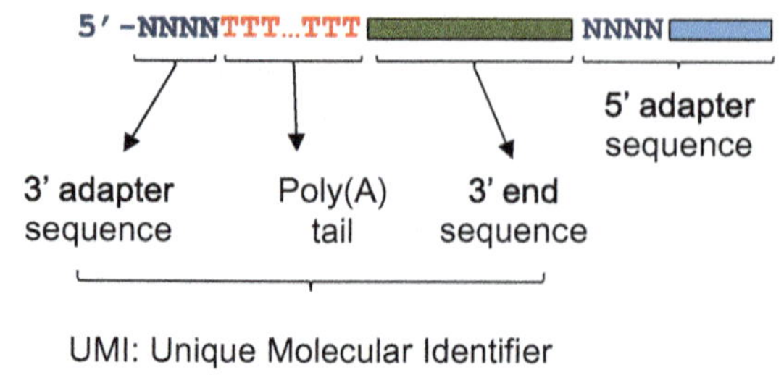

Fig. 2 Schematic showing different parts of a typical raw read generated by 3′READS+. N, random nucleotide from adapters; T, from poly(A) tail. 5′ adapter sequence is also shown

6. Elute DNA from SPRI beads by resuspending the beads in 25 μl of 10 mM Tris–HCl pH 8.0 and incubate for 5 min at room temperature. Place the tubes on a magnetic stand. After the solution is clear, transfer the supernatant to a new tube.
7. If necessary, a second round of purification to eliminate short PCR products will be carried out as follows: Add 22.5 μl of fresh SPRI beads to 25 μl of the eluted DNA. Mix well by pipetting up and down a few times, and incubate at room temperature for 5 min. Place the tubes on a magnetic stand. After the solution is clear, discard the supernatant, which mostly contains small PCR products, and keep the beads. Repeat **steps 4–6**.

3.11 Library Quantification and Normalization

1. Take 1 μl of the eluted libraries for Bioanalyzer analysis or 2 μl for TapeStation analysis. Most of the PCR products should be between 200 and 500 bp.
2. Use Qubit to measure the mass concentration of library and use the average molecular size reported by Bioanalyzer or TapeStation to calculate molar concentration for each library.
3. Mix the libraries so that different libraries have approximately the same molar concentration in the mixture.
4. The mixed libraries are ready for sequencing.
5. A typical 3′READS+ read is illustrated in Fig. 2.

3.12 Mapping Cleavage Sites in Genome

1. Remove the sequence corresponding to the 5′ adapter from raw 3′READS+ reads using the cutadapt program (http://journal.embnet.org/index.php/embnetjournal/article/view/200). The following arguments are used: -a NNNNTGGAATTCTCGGGTGCCAAGG --overlap 7 -m 30, where the sequence after "-a" corresponds to the 5′ adapter sequence, "--overlap 7" indicates that at least 7 nt of overlap (including 4 Ns) is required to identify the adapter sequence. "-m 30" indicates that reads shorter than 30 nt after trimming will be discarded.

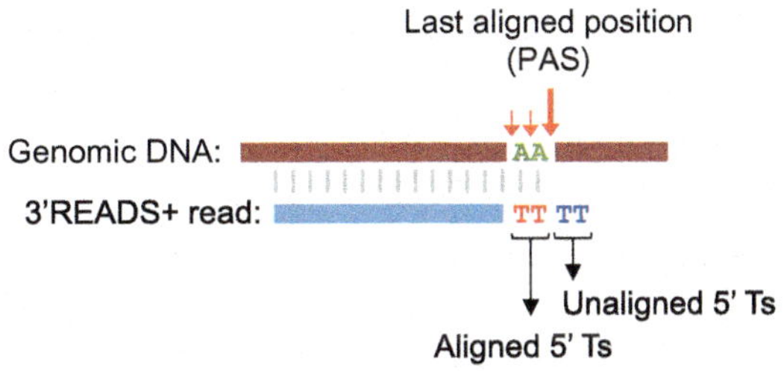

Fig. 3 Schematic showing alignment of a PAS-supporting (PASS) read to a genomic sequence. Aligned and unaligned 5′Ts are indicated. The number of unaligned 5′Ts is used to call a PAS. The last aligned position is the PAS position (indicated by a *big arrow*). Note the real cleavage position can be before any As immediately upstream of the last aligned position (indicated by *small arrows*)

2. Remove the 5′ random nucleotides (from the 3′ adapter). The sequence of the random nucleotides is saved for later analysis as part of Unique Molecular Identifier (UMI, Fig. 2).
3. Remove the 5′ Ts in reads, allowing up to one non-T base in the 5′ Ts. The length of 5′ Ts is saved for later analysis as evidence of poly(A) tail and as part of UMI (Fig. 2). If the trimmed 5′ Ts contains a non-T base, its position and identity are recorded. Reads shorter than 23 nt after this step are filtered out.
4. Map the reads to the genome using bowtie2 in local mode [14]. Only reads with a mapping quality score (MAPQ) ≥10 and no soft clipping at the 5′ end are used for further analysis. The genomic position of the 5′ end of the read is considered the last aligned position (Fig. 3). If the trimmed 5′ Ts contains a non-T base and the non-T base can match the genomic sequence, the position of the non-T base is considered the last aligned position and the number of trimmed Ts is reduced accordingly.
5. The number of trimmed 5′Ts of each read from **steps 2** and **4** are compared with the number of As downstream of the last aligned position in the genome. The reads with ≥2 unmappable 5′Ts after this process are considered PAS-supporting (PASS) reads. The last aligned position is considered PAS (Fig. 3). Note when the last aligned position is preceded by As, the real cleavage location cannot be resolved.

3.13 Clustering of Cleavage Sites

1. Because cleavage and polyadenylation can take place at multiple places in a small window of sequence, PASs are clustered. We first progressively cluster together PASs located within 24 nt from one another. If a cluster size is ≤24 nt, the PAS with the greatest number of PASS reads is chosen as the representative position for the cluster; if a cluster is wider than 24 nt,

we identify the PAS with the greatest number of PASS reads to represent all PASs within 24 nt from the site. The remaining PASs are then processed using the same strategy until no PASs are left.

3.14 Associate PASs to Genes

1. PASs are assigned to gene features using the RefSeq data from NCBI, including 3′UTRs, CDS, 5′UTRs, introns, upstream antisense region, etc.
2. The 3′ ends of gene models defined by RefSeq can be extended using strand-specific RNA-seq data to increase the probability for PASs to be assigned to genes.
3. While expression of each APA isoform can be examined separately, all PASS reads mapped to the genic region can be summed up to represent the expression level of the gene (*see* **Note 3**).
4. Alternative cleavage and polyadenylation can be studied by comparing the expression of APA isoforms of the same genes across samples using statistical methods, such as Fisher's exact test, DEXSeq [15], or Significance Analysis of Alternative Polyadenylation (SAAP) [16].

4 Notes

1. In this section, it is assumed that several libraries are processed together in one batch. Sample-to-sample variability increases when multiple samples are handled at the same time. To ensure library quality, the number of samples should be not more than 16.
2. Sensitivity of 3′READS+. 3′READS+ requires ≥100 ng total RNA as input, a vast improvement over the 3′READS method we previously developed, which required 25 μg of total RNA [10]. This is due to better poly(A)+ RNA capture and much improved ligation steps. Input less than 100 ng may also work, but may compromise the library quality. For example, the amount of cDNAs containing short inserts may increase. This can be partially addressed by using multiple rounds of cDNA purification by SPRI beads.
3. Quantitative analysis of transcripts. In addition to read number, the 3′READS+ data can be analyzed using unique molecular identifiers (UMI) based on (1) number of 5′Ts, (2) random nucleotides (from 3′ adapter), (3) length of sequence aligned to genome. UMI measures the abundance of RNA fragments. Its use can address problem of over-amplification of RT products by PCR.
4. Ligation bias. The random four Ns in 5′ and 3′ adapters can help mitigate the bias of RNA ligases [17, 18]. In addition, the

four Ns in the 3′ adapter can (1) facilitate cluster calling by the Illumina sequencing software, which uses the first 4 nucleotides to distinguish clusters; and (2) be used as part of UMI.

References

1. Colgan DF, Manley JL (1997) Mechanism and regulation of mRNA polyadenylation. Genes Dev 11(21):2755–2766
2. Tian B et al (2005) A large-scale analysis of mRNA polyadenylation of human and mouse genes. Nucleic Acids Res 33(1):201–212
3. Tian B, Manley JL (2016) Alternative polyadenylation of mRNA precursors. Nat Rev Mol Cell Biol 18(1):18–30
4. Hrdlickova R, Toloue M, Tian B (2016) RNA-seq methods for transcriptome analysis. Wiley Interdiscip Rev RNA 8(1)
5. Velculescu VE et al (1995) Serial analysis of gene expression. Science 270(5235):484–487
6. Shepard PJ et al (2011) Complex and dynamic landscape of RNA polyadenylation revealed by PAS-Seq. RNA 17(4):761–772
7. Derti A et al (2012) A quantitative atlas of polyadenylation in five mammals. Genome Res 22(6):1173–1183
8. Fu Y et al (2011) Differential genome-wide profiling of tandem 3′ UTRs among human breast cancer and normal cells by high-throughput sequencing. Genome Res 21(5):741–747
9. Jan CH et al (2011) Formation, regulation and evolution of *Caenorhabditis elegans* 3′UTRs. Nature 469(7328):97–101
10. Hoque M et al (2013) Analysis of alternative cleavage and polyadenylation by 3′ region extraction and deep sequencing. Nat Methods 10(2):133–139
11. Nam DK et al (2002) Oligo(dT) primer generates a high frequency of truncated cDNAs through internal poly(A) priming during reverse transcription. Proc Natl Acad Sci U S A 99(9):6152–6156
12. Lee JY, Park JY, Tian B (2008) Identification of mRNA polyadenylation sites in genomes using cDNA sequences, expressed sequence tags, and trace. Methods Mol Biol 419:23–37
13. Zheng D, Liu X, Tian B (2016) 3′READS+, a sensitive and accurate method for 3′ end sequencing of polyadenylated RNA. RNA 22(10):1631–1639
14. Langmead B, Salzberg SL (2012) Fast gapped-read alignment with bowtie 2. Nat Methods 9(4):357–359
15. Anders S, Reyes A, Huber W (2012) Detecting differential usage of exons from RNA-seq data. Genome Res 22(10):2008–2017
16. Li W et al (2015) Systematic profiling of poly(A)+ transcripts modulated by core 3′ end processing and splicing factors reveals regulatory rules of alternative cleavage and polyadenylation. PLoS Genet 11(4):e1005166
17. Jayaprakash AD et al (2011) Identification and remediation of biases in the activity of RNA ligases in small-RNA deep sequencing. Nucleic Acids Res 39(21):e141
18. Zhuang F et al (2012) Structural bias in T4 RNA ligase-mediated 3′-adapter ligation. Nucleic Acids Res 40(7):e54

Chapter 7

Comprehensive Identification of mRNA Polyadenylation Sites by PAPERCLIP

Hun-Way Hwang and Robert B. Darnell

Abstract

We discuss a newly developed method to profile mRNA polyadenylation (pA) sites in an unbiased manner, PAPERCLIP (*P*oly(*A*) binding *P*rotein-mediated mRNA 3′*E*nd *R*etrieval by *C*ross*L*inking *I*mmuno*P*recipitation). Based on the well-established CLIP (crosslinking immunoprecipitation) technique, PAPERCLIP utilizes the poly(A) binding protein (PABP) as a biological filter to selectively retrieve mRNA 3′ end fragments by immunoprecipitation from ultraviolet (UV) irradiated tissues or cultured cells. The mRNA fragments are subsequently extracted from the immunoprecipitated PABP:RNA complexes to generate a cDNA library, which goes through two rounds of purification before the final amplification by real-time polymerase chain reaction (PCR). The amplified cDNA library can then be read out by high-throughput sequencing to generate a transcriptomic profile and comprehensive alternative poly(A) (APA) site map from intact tissue or cultured cells.

Key words mRNA polyadenylation, Crosslinking immunoprecipitation, Poly(A)-binding protein

1 Introduction

A simple and commonly used method to generate pA site annotations is to construct a high-throughput sequencing library by reverse transcribing mRNAs from the poly(A) tails utilizing oligo(dT) primers [1]. Nevertheless, it is also recognized that internal priming events, which are caused by annealing of oligo(dT) primers to internal adenine-rich mRNA sequences, can represent 30% of the entire library constructed by this simple approach [2]. These internal priming events confound the interpretation of sequencing results and introduce errors to the resulting pA site annotations. Therefore, subsequent in silico filtering is usually implemented to remove internal priming events [3]. Although it improves the accuracy of pA site annotations in general, in silico filtering could bias the annotations and may eliminate authentic pA sites [2, 4].

Yongsheng Shi (ed.), *mRNA Processing: Methods and Protocols*, Methods in Molecular Biology, vol. 1648,
DOI 10.1007/978-1-4939-7204-3_7, © Springer Science+Business Media LLC 2017

As accumulating evidence suggests an important role of alternative polyadenylation (APA) in basic cellular functions and in human disease [5], we sought to develop a more robust biochemical pA site mapping method that is highly selective for the mRNA 3′ ends. We developed PAPERCLIP by building upon the CLIP technique, which generates precise genome-wide interaction maps by capturing direct protein-RNA interactions in situ by utilizing UV-induced covalent crosslinking of such interactions when they are close (within ~1 Angstrom), followed by stringent immunoprecipitation to eliminate nonspecific protein-RNA interactions [6]. In PAPERCLIP, the mRNA 3′ end fragments containing the poly(A) tail are selectively purified through their direct interaction with PABP, which has high specificity for mRNA poly(A) tails [7]. We demonstrated that PAPERCLIP minimizes the internal priming problem without utilizing in silico filtering and allows unbiased discovery of mRNA 3′ ends [8]. The ability of PAPERCLIP to annotate all poly(A) sites in a tissue has the ability to compliment RNAseq mRNA quantitation methods, and may be particularly useful in the future if the approach can be applied to individual cell types [9] where single cell-type RNA measurements are difficult. Notably, the PAPERCLIP protocol can be easily modified to perform CLIP experiments for other proteins of interest and we showed that combining PAPERCLIP and Argonaute CLIP results in the discovery of new regulatory relationships between mRNAs and miRNAs [8]. The methodological synergy between PAPERCLIP and other CLIP experiments thus makes it an especially attractive method for identifying different mRNA APA isoforms and elucidating their biological functions.

2 Materials

2.1 Ultraviolet Crosslinking of Cultured Cells

1. 1× phosphate-buffered saline (tissue culture grade, Mg^{2+}/Ca^{2+}-free).
2. Plastic tray.
3. UV crosslinker.
4. Refrigerated Microcentrifuge.

2.2 Immunoprecipitation

1. Magnetic Rack.
2. Dynabeads Protein G.
3. Antibody Binding Buffer: 1× phosphate-buffered saline (prepare from tissue culture grade Mg^{2+}/Ca^{2+}-free 10× PBS), 0.02% Tween-20.
4. Mouse monoclonal antibody against PABP, clone 10E10.
5. 1× TS Buffer: 1× phosphate-buffered saline (prepare from tissue culture grade Mg^{2+}/Ca^{2+}-free 10× PBS), 0.1% SDS, 1.0% Triton X-100.

6. RQ1 DNase.
7. Incubator/Shaker.
8. RNase A (Molecular Biology Grade).
9. High RNase A solution: 1:100 dilution of RNase A in 1× TS Buffer.
10. Low RNase A solution: 1:100,000 dilution of RNase A in 1× TS Buffer.
11. Refrigerated Microcentrifuge.
12. 2× TS Buffer: 2× phosphate-buffered saline (prepare from tissue culture grade Mg^{2+}/Ca^{2+}-free 10× PBS), 0.1% SDS, 1.0% Triton X-100.
13. 1× PNK Wash Buffer: 50 mM Tris–HCl, pH 7.4, 10 mM $MgCl_2$, 0.5% NP-40.

2.3 5′ Labeling of Immunoprecipitated RNA

1. Fast AP.
2. 1× PNK + EGTA Wash Buffer: 50 mM Tris–HCl, pH 7.4, 20 mM EGTA, 0.5% NP-40.
3. T4 PNK.
4. ^{32}P-γ-ATP.

2.4 SDS-PAGE and Nitrocellulose Transfer

1. NuPAGE 4× LDS Sample Buffer.
2. 1 M dithiothreitol (DTT).
3. NuPAGE 10% Bis-Tris Midi Gel.
4. NuPAGE 20× MOPS SDS Running Buffer.
5. Full-Range Rainbow Molecular Weight Markers.
6. Nitrocellulose membrane.
7. NuPAGE 20× Transfer Buffer.
8. Autorad Markers.

2.5 RNA Isolation and Purification

1. 1× PK Buffer: 100 mM Tris–HCl, pH 7.5, 50 mM NaCl, 10 mM EDTA.
2. Working Proteinase K Solution: Dilute Proteinase K solution (20 mg/mL, PCR grade) 1:5 in 1× PK Buffer. (Must be made fresh.)
3. Working Proteinase K Solution in 1× PK Buffer/7 M urea: Add 0.42 g Urea to 1 mL Working Proteinase K Solution. Vortex until urea is complete dissolved. (Must be made fresh.)
4. RNA phenol.
5. Chloroform.
6. Ultrapure glycogen.
7. 3 M NaOAc, pH 5.2.
8. 1:1 mix of ethanol and isopropanol.

2.6 cDNA Synthesis and Purification I

1. 50× Denhardt's Solution.
2. Nuclease-free water (RT-PCR grade).
3. 75% ethanol.
4. 8.2 mM dATP/dCTP/dGTP (prepare from 100 mM stock).
5. 8.2 mM BrdUTP (prepare from powder).
6. 5× RT Buffer/0.1MDTT/SuperScript III.
7. Indexed RT primers (Order with PAGE purification):

   RT-1 (CGAT): /5Phos/DDDATCGNNNNNNNAGATCGGAAGAGCGTCGT/iSp18/CACTCA/iSp18/CAA

   GCAGAAGACGGCATACGAGATTTTTTTTTTTTTTTTTTTTVN

   RT-2 (TAGC): /5Phos/DDDGCTANNNNNNNAGATCGGAAGAGCGTCGT/iSp18/CACTCA/iSp18/CAA

   GCAGAAGACGGCATACGAGATTTTTTTTTTTTTTTTTTTTVN

   RT-3 (CTAG):

   /5Phos/DDDGACTNNNNNNNAGATCGGAAGAGCGTCGT/iSp18/CACTCA/iSp18/CAA

   GCAGAAGACGGCATACGAGATTTTTTTTTTTTTTTTTTTTVN

   RT-4 (GATC):

   /5Phos/DDDAGTCNNNNNNNAGATCGGAAGAGCGTCGT/iSp18/CACTCA/iSp18/CAA

   GCAGAAGACGGCATACGAGATTTTTTTTTTTTTTTTTTTTVN
8. Anti-BrdU, clone IIB5.
9. 4× IP Buffer: 1.2× SSPE, 4 mM EDTA, 0.2% Tween-20.
10. 2× IP Buffer: 1:2 dilution of 4× IP Buffer.
11. 1× IP Buffer: 1:4 dilution of 4× IP Buffer.
12. RNase H (make 2 U/μL working solution).
13. Microspin G-25 column.
14. Nelson Low Salt Buffer: 15 mM Tris–HCl, pH 7.5, 5 mM EDTA.
15. Nelson Stringent Buffer: 15 mM Tris–HCl, pH 7.5, 5 mM EDTA, 2.5 mM EGTA, 120 mM NaCl, 25 mM KCl, 1% Triton X-100, 1% Sodium deoxycholate, 0.1% SDS.

2.7 cDNA Purification II and Library Construction

1. CircLigase Wash Buffer: 33 mM Tris-Acetate, 66 mM KCl, pH 7.8.
2. CircLigase II (includes 10× Reaction Buffer, 5 M Betaine and 50 mM $MnCl_2$)
3. Phusion Wash Buffer: 50 mM Tris–HCl, pH 8.0.

4. Phusion DNA Polymerase.
5. 10 mM dNTPs.
6. 50× SYBR Green I.
7. DP5-PE primer (Order with PAGE purification): 5′-AATGATACGGCGACCACCGAGATCTACACTCTTTCCCTACACGACGCTCTTCCGATCT.
8. DP3-PAT primer: 5′-CAAGCAGAAGACGGCATA.
9. 0.2 mL PCR tube compatible with real-time PCR.
10. Thermocycler for real-time PCR
11. Agencourt AMPure XP Beads.
12. TapeStation.

3 Methods

The entire PAPERCLIP protocol can be completed in 5 days with a typical 8-h workday schedule:

Day 1: Ultraviolet Crosslinking of Cultured Cells (Subheading 3.1), Immunoprecipitation (Subheading 3.2), and 5′ labeling of immunoprecipitated RNA (Subheading 3.3).

Day 2: SDS-PAGE and Nitrocellulose Transfer (Subheading 3.4).

Day 3: RNA Isolation and Purification (Subheading 3.5).

Day 4: cDNA Synthesis and Purification I (Subheading 3.6).

Day 5: cDNA Purification II and Library Construction (Subheading 3.7).

If desired, an alternative schedule with possible 9 ~ 10-h of workload for some days to speed up the process:

Day 1: Ultraviolet Crosslinking of Cultured Cells (Subheading 3.1), Immunoprecipitation (Subheading 3.2), 5′ labeling of immunoprecipitated RNA (Subheading 3.3), SDS-PAGE and Nitrocellulose Transfer (Subheading 3.4).

Day 2: RNA Isolation and Purification (Subheading 3.5).

Day 3: cDNA Synthesis and Purification I (Subheading 3.6), cDNA Purification II, and Library Construction (Subheading 3.7).

3.1 Ultraviolet Crosslinking of Cultured Cells

1. *UV crosslinking*: Remove growth media from adherent cells grown in 10 cm dishes in culture. Wash once with ice-cold 1× PBS and add a thin layer (for example, 4 mL) of ice-cold 1× PBS for each dish (*see* **Note 1**). To facilitate processing, use a tray filled with ice to hold multiple dishes on ice. Put the tray in UV crosslinker, remove the lids, and irradiate the dishes three times at 200 mJ/cm^2 (*see* **Note 2**). If possible, rotate the tray between each irradiation to ensure an even distribution of crosslinking. *See* **Note 3** about using mouse tissue for crosslinking.

2. *Post-crosslinking processing*: Scrape the cells from the dishes in the original thin layer of ice-cold 1× PBS. Pool the cell suspension and divide them into multiple microfuge tubes such that each microfuge tube contains sufficient materials to make one PAPERCLIP library (*see* **Note 4**). Spin the microfuge tubes to pellet the cells at 4 °C. Remove the supernatant and proceed to immunoprecipitation. Alternatively, the cell pellets can be frozen at −80 °C until use.

3.2 Immunoprecipitation

1. *Magnetic bead preparation*: Prepare beads as a master mix for all groups (*see* **Note 5**). Wash the beads three times with Antibody Binding Buffer using a magnetic rack. When the washing is finished, resuspend the beads in the appropriate amount of Antibody Binding Buffer containing the 10E10 anti-PABP (*see* **Note 6**) and rotate the beads for at least 30 min at room temperature to bind the antibodies. Proceed to prepare crosslinked lysates.
2. *Prepare crosslinked lysates*: Resuspend each tube of crosslinked tissue in appropriate amount of 1× TS Buffer (*see* **Note 7**) and incubate on ice with occasional vortexing for 5 min to lyse. Add 30 ~ 40 μL of RQ1 DNase to each tube and incubate the tubes at 37 °C for 5 min and shake at 1000 rpm on an incubator/shaker (*see* **Note 8**). Put the tubes back on ice. For the High-RNase group, add 10 ~ 20 μL of High RNase A solution per 1 mL lysate. For the Low-RNase group, add 10 μL of Low RNase A solution per 1 mL lysate. Incubate the tubes at 37 °C for 5 min and shake at 1000 rpm on an incubator/shaker. Clear the lysates by spinning the tubes in a prechilled microcentrifuge at ~20,000 × *g* for 10 min at 4 °C. Transfer the supernatant to new tubes and leave the tubes on ice.
3. *IP*: On a magnetic rack, wash the antibody-bound beads three times with 1× TS Buffer and divide them into a number of tubes that corresponds to the number of IP groups in the last wash. Remove the last wash solution from the beads and add the cleared lysate for each tube. Rotate the tubes for 2 h at 4 °C.
4. *Post-IP washes*: Wash the beads on a magnet rack in the following order: twice with 1× TS Buffer, twice with 2× TS Buffer, and twice with 1× PNK Wash Buffer (*see* **Note 9**).

3.3 5′ Labeling of Immunoprecipitated RNA

1. *Alkaline phosphatase treatment*: Make a master mix for all the tubes. For each reaction, add 51.1 μL water, 6 μL 10× Fast AP Buffer, 0.6 μL 10% Tween-20, and 2.3 μL Fast AP (*see* **Note 10**). Remove the last 1× PNK Wash from the beads and add 60 μL per tube on a magnet rack. Remove the tubes from the magnet rack and mix the beads and the solution by brief and gentle vortexing or finger tapping. Incubate the tubes at 37 °C

for 20 min and shake at 1200 rpm for 15 s every 90 s on an incubator/shaker. When the incubation is complete, wash the beads in the following order: once with 1× PNK Wash Buffer, once with 1× PNK + EGTA Wash Buffer, and twice with 1× PNK Wash Buffer.

2. *PNK 5′ labeling*: Make a master mix for all the tubes. For each reaction, add 49.5 μL water, 6 μL 10× PNK Buffer, 3 μL T4 PNK, and 1.5 μL ^{32}P-γ-ATP (*see* **Note 11**). Remove the last 1× PNK Wash from the beads and add 60 μL per tube on a magnet rack. Remove the tubes from the magnet rack and mix the beads and the solution by brief and gentle vortexing. Incubate the tubes at 37 °C for 20 min and shake at 1200 rpm for 15 s every 90 s on an incubator/shaker. When the incubation is complete, wash the beads three times with 1× PNK Wash Buffer. Leave the beads in the last wash and store at 4 °C overnight.

3.4 SDS-PAGE and Nitrocellulose Transfer

1. *Elution*: Make a master mix for all the tubes. For each reaction, add 30 μL 4× LDS Sample Buffer, 27 μL 1× PNK Wash Buffer, 3 μL 1 M DTT. Remove the last 1× PNK Wash from the beads and add 60 μL per tube on a magnet rack. Remove the tubes from the magnet rack and mix the beads and the solution by brief and gentle vortexing. Incubate the tubes at 70 °C for 10 min and shake at 1200 rpm on an incubator/shaker. When the incubation is complete, put the tubes back on a magnet and transfer the solution to a new microfuge tube. Discard the original tubes containing beads.
2. *SDS-PAGE*: Load one tube per two wells (30 μL/well) of a 20-well 10% gel. Run the gel at 200 V in the cold room with 1× MOPS buffer for about 1.5 h (*see* **Note 12**).
3. *Transfer*: Transfer the RNA:protein complex from the gel to a pure nitrocellulose membrane using a Novex wet transfer apparatus in 1× NuPAGE Transfer Buffer supplemented with 10% methanol. Perform transfer in a cold room at 30 V for 75 ~ 90 min.
4. *Film exposure*: When the transfer is complete, rinse the nitrocellulose membrane briefly in 1× PBS and gently blot on Kimwipes to remove excess buffer. Enclose the membrane in plastic wrap and expose it to a film in a cassette at −80 °C (*see* **Notes 13** and **14** and Fig. 1).

3.5 RNA Isolation and Purification

1. *Cutting the nitrocellulose membrane*: Make a working proteinase K solution in 1× PK Buffer for all groups (*see* **Note 15**). Incubate the solution at 37 °C for 10 ~ 20 min to remove RNase. Using the film as a guide, label the corners of the desired area on the nitrocellulose membrane (*see* Fig. 1). When the labeling is complete, first cut out the entire area of

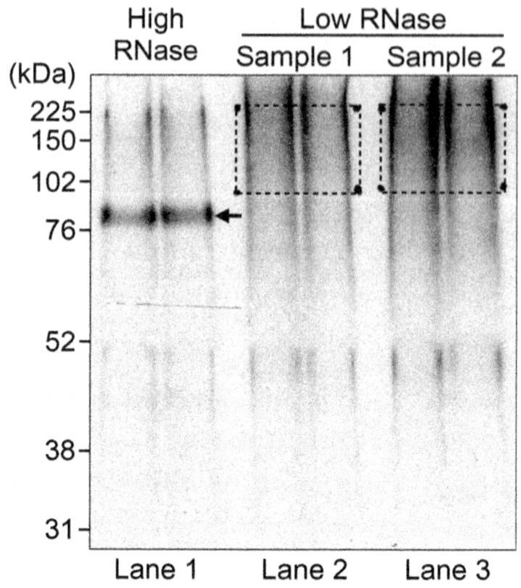

Fig. 1 An autoradiogram from a two-sample PAPERCLIP experiment. A clear band can be seen from the High RNase group (Lane 1, *left*). (The molecular weight of PABP is about 75 kDa and the *black arrow* denotes the PABP:RNA complex.) In contrast, a desired smear pattern is observed for the Low RNase groups (Lanes 2 and 3, *middle* and *right*). The *dashed lines* show the margins of the nitrocellulose membranes used for RNA extraction and subsequent sequencing library construction. The *dot labels* on the original film to mark the corners of the desired areas on the nitrocellulose membrane can also be seen

nitrocellulose membrane using a clean scalpel blade and then further cut it into multiple small pieces on a filter paper. Use the scalpel blade to collect the pieces into a microfuge tube. Repeat the cut-collection process for each sample and change blades between samples.

2. *Proteinase K digestion*: Add 200 μL working proteinase K solution to each microfuge tube containing the nitrocellulose membrane pieces and incubate the tubes at 37 °C for 20 min and shake at 1200 rpm on an incubator/shaker.
3. *Proteinase K/7 M urea digestion*: Add 200 μL working proteinase K solution in 1× PK Buffer/7 M Urea to each microfuge tube containing the nitrocellulose membrane pieces and incubate the tubes at 37 °C for 20 min and shake at 1200 rpm on an incubator/shaker.
4. *RNA extraction*: Add 400 μL RNA phenol and 130 μL chloroform to each microfuge tube. Vortex vigorously for 20 s. Incubate the tubes at 37 °C for 20 min and shake at 1200 rpm on an incubator/shaker. Spin all tubes in a prechilled microcentrifuge at full speed (>20,000 *g*) for 5 min at 4 °C. Transfer aqueous phase (~500 μL) to new microfuge tubes. Add the following to each tube: 1 μL glycogen, 50 μL 3 M NaOAc, 1 mL ethanol-isopropanol mixture. Thoroughly mix and precipitate overnight at −20 °C.

3.6 cDNA Synthesis and Purification I

1. *Magnetic bead preparation, blocking*: Prepare beads as a master mix for all groups (*see* **Note 16**). Wash the beads three times with Antibody Binding Buffer using a magnetic rack. When the washing is finished, resuspend the beads in the appropriate amount of Antibody Binding Buffer supplemented with 50× Denhardt's Solution (*see* **Note 17**) and rotate the beads for at least 45 min at room temperature to block nonspecific binding. Proceed to precipitate RNA.
2. *RNA precipitation*: Spin all microfuge tubes that are precipitated overnight in a prechilled microcentrifuge at full speed (>20,000 *g*) for 20 min at 4 °C to pellet the RNA (*see* **Note 18**). Remove the supernatant and wash the pellets by adding 200 μL 75% ethanol and spinning at full speed for 10 min at 4 °C. Repeat ethanol wash once and dry the pellet (*see* **Note 19**). Add 8 μL nuclease-free water to the pellet. Incubate the tube at 65 °C for 5 min. After incubation, do a quick spin-down and put the microfuge tubes on ice (*see* **Note 20**). Carefully resuspend the RNA by gently pipetting up and down a few times before transferring it to 0.2 mL PCR tubes. Pool 0.5 ~ 1 μL from all groups to an extra tube as the minus RT control. Bring the volume of minus RT control to 8 μL with nuclease-free water.
3. *Prepare for reverse transcription*: First, in a microfuge tube, prepare RT-PCR Master Mix I for all groups including the minus RT tube: add 4 μL 5× RT Buffer, 1 μL 8.2 mM dATP, 1 μL 8.2 mM dCTP, 1 μL 8.2 mM dGTP, and 1 μL 8.2 mM Br-dUTP per group. Second, in a PCR tube, prepare RT-PCR Master Mix IIA for all groups (except for the minus RT group): add 1 μL 0.1 M DTT, 1 μL RNasin Plus, and 1 μL SuperScript III per group. Last, in a PCR tube, prepare RT-PCR Master Mix IIB for the minus RT group: add 1 μL 0.1 M DTT, 1 μL RNasin Plus, and 1 μL nuclease-free water. Start and hold the thermocycler at 75 °C.
4. *Reverse transcription*: Add 8 μL of Mix I and 1 μL of 25 μM indexed RT primer to each sample PCR tube including the minus RT group (*see* **Note 21**). Transfer the sample PCR tubes to the thermocycler heat block. Denature at 75 °C for 3 min. Ramp down to 48 °C and hold. Transfer RT-PCR Master Mix IIA and IIB to the thermocycler heat block and incubate briefly. Add 3 μL of Mix II to each sample. Continue with reverse transcription: 45 min at 48 °C, 15 min at 55 °C, 5 min at 85 °C and hold at 4 °C. Proceed to the next step when reverse transcription is underway.
5. *Magnetic bead preparation, antibody binding*: On a magnetic rack, wash the blocked beads three times with Antibody Binding Buffer. Prepare a master mix binding solution for all groups: add 37.5 μL Antibody Binding Buffer, 5 μL 50× Denhardt's Solution and 5 μL (5 μg) anti-BrdU antibody for

each group. Resuspend all the beads in the binding solution and rotate the beads for at least 45 min at room temperature to bind the antibodies (*see* **Note 22**).

6. *Post-RT cleanup*: When reverse transcription is complete, add 1 μL of RNase H to each PCR tube of RT reaction. Incubate at 37 °C for 20 min on the thermocycler. When the RNase H digestion is complete, transfer the PCR tubes back on ice. Add 10 μL nuclease-free water and do a quick spin down (*see* **Note 23**). Prepare the G-25 columns following the manufacturer's instructions. Spin each RT reaction through a G-25 column into a new microfuge tube at 750 × *g* for 2 min to remove excess BrdUTP.
7. *cDNA purification: immunoprecipitation I*: Measure the volume of each cleared RT reaction by pipet and bring the volume up to 40 μL by adding nuclease-free water. Then add 50 μL 2× IP Buffer and 10 μL 50× Denhardt's Solution for a total volume of 100 μL (*see* **Note 24**). Denature the samples at 70 °C for 5 min and equilibrate to room temperature. On a magnetic rack, wash the antibody-bound beads three times with 1× IP Buffer. Store half of the washed beads at 4 °C and split the remaining half by the number of groups. Remove wash buffer from the beads and add the denatured sample. Tap to mix the sample and the beads. Rotate for 45 min at room temperature.
8. *cDNA purification: post-IP wash I*: Wash beads on a magnet rack in the following order: once with 1× IP Buffer plus 5× Denhardt's Solution, twice with Nelson Low Salt Buffer plus 1× Denhardt's Solution, twice with Nelson Stringent Buffer plus 1× Denhardt's Solution, and twice with 1× IP Buffer (*see* **Note 25**). For each wash, the tubes are rotated in 1 mL of the corresponding wash buffer for 3 min at room temperature (*see* **Note 26**).
9. *cDNA purification: heat elution*: Add 90 μL Elution Buffer to each tube. Incubate the tubes at 98 °C for 1 min and shake at 1200 rpm on an incubator/shaker. When the incubation is complete, immediately place the tubes on a magnet. Collect eluate and add 10 μL 50× Denhardt's Solution for a total volume of 100 μL. Store the samples overnight at 4 °C.

3.7 cDNA Purification II and Library Construction

1. *cDNA purification: immunoprecipitation II*: Denature the samples at 70 °C for 5 min and equilibrate to room temperature. Split the remaining beads by the number of groups. Remove wash buffer from the beads and add the denatured sample. Tap to mix the sample and the beads. Rotate for 45 min at room temperature (*see* **Note 27**).
2. *cDNA purification: post-IP wash II*: Wash beads on a magnet rack in the following order: once with 1× IP Buffer plus 5× Denhardt's Solution, twice with Nelson Low Salt Buffer plus

1× Denhardt's Solution, twice with Nelson Stringent Buffer plus 1× Denhardt's Solution, and twice with CircLigase Wash Buffer. For each wash, the tubes are rotated in 1 mL of the corresponding wash buffer for 3 min at room temperature.

3. *cDNA circularization*: Prepare a master mix for the Circligase Reaction: add 12 μL nuclease-free water, 2 μL 10× CircLigase Reaction Buffer, 4 μL 5 M Betaine, 1 μL 50 mM $MnCl_2$, and 1 μL CircLigase II per group. Add 20 μL master mix to each tube. Incubate the tubes at 60 °C for 1 h and shake at 1300 rpm for 15 s every 30 s on an incubator/shaker. When the incubation is complete, wash the beads on a magnet rack in the following order: twice with Nelson Low Salt Buffer, twice with Nelson Stringent Buffer, and twice with Phusion Wash Buffer. For each wash, the tubes are rotated in 1 mL of the corresponding wash buffer for 3 min at room temperature.
4. *cDNA elution and PCR amplification*: Prepare Master Mix I for all groups: add 37 μL nuclease-free water, 10 μL 5× Phusion HF Buffer, and 1 μL 10 mM dNTPs per group (*see* **Note 28**). Prepare Master Mix II for all groups: add 0.5 μL DP5-PE (20 μM), 0.5 μL DP3-PAT (20 μM), and 0.5 μL Phusion DNA Polymerase per group. Make 50× SYBR Green I Solution by adding 1 μL stock solution (10,000×) to 199 μL Phusion Wash Buffer. Add 48 μL Mix I to washed beads. Incubate the tubes at 98 °C for 1 min and shake at 1200 rpm on an incubator/shaker to elute the cDNAs. When the incubation is complete, immediately place the tubes on a magnet. Transfer the eluate into PCR tubes (*see* **Note 29**). Add 1.5 μL Mix II and 0.5 μL 50× SYBR Green I Solution to each tube. Start and hold the thermocycler at 98 °C. Transfer the PCR tubes to the thermocycler and start real-time PCR. The initial denaturation is 98 °C for 30 s. The cycling parameters are: 10 s at 98 °C, 15 s at 60 °C, and 20 s at 72 °C. Remove the reaction tube when the fluorescence signal reaches 250 ~ 500 (*see* **Note 30** and Fig. 2).
5. *Post-PCR processing: library purification and quantitation*: Purify PCR product using Agencourt AMPure XP Beads following manufacturer's instructions (use 90 μL beads per 50 μL PCR reaction). Quantitate the concentration of each library using TapeStation.

4 Notes

1. To harvest suspension cells, spin down the cells to remove growth media, wash once with ice-cold 1× PBS. Resuspend cells in ice-cold 1× PBS at the density of 5 ~ 10 million cells/mL. Crosslink the suspension in 10 cm dishes the same way as for adherent cells.

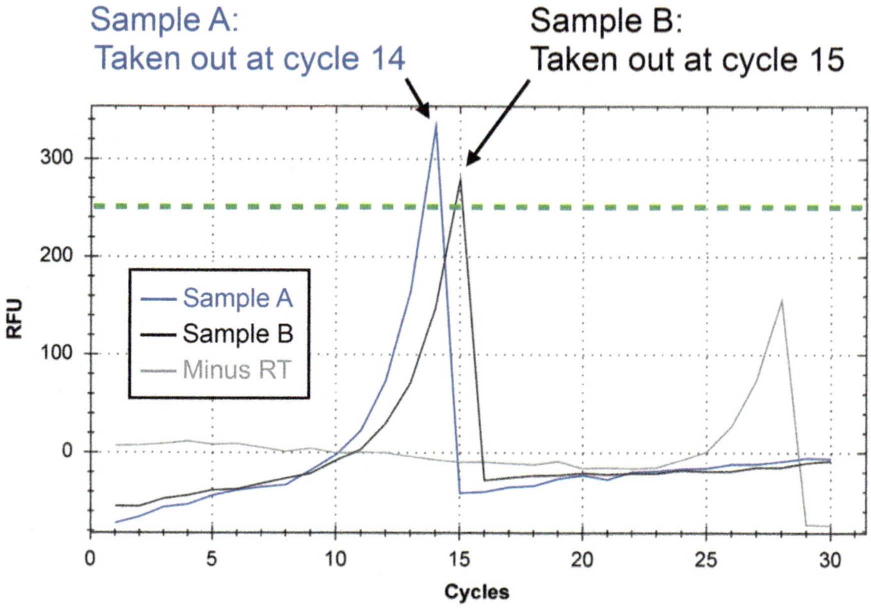

Fig. 2 A diagram showing the PAPERCLIP library amplification results on a Bio-Rad CFX96 Touch™ Real-Time PCR Detection System. The X-axis shows the number of PCR cycles and the Y-axis shows the fluorescence signal (RFU, relative fluorescence units). Each trace represents individual samples. The *dashed line* at RFU 250 (which was not present in the original diagram and was manually added for illustrative purposes) denotes the minimum threshold for library amplification. Both Samples A and B were taken out of the thermocycler as soon as they passed 250 in RFU (*arrows*) and resulted in a sharp dip in fluorescence signal in the subsequent cycle in the diagram. This diagram also shows amplification of the minus RT group at a much later cycle (cycle 28), which usually comes from self-ligated residual RT primers

2. The setting for UV crosslinking is based on Stratalinker. Optimization might be necessary if a different crosslinker is used.
3. To enhance crosslinking efficiency for mouse tissue, the tissue needs to be triturated a few times with an 18-gauge needle and syringe and irradiated at a higher intensity (400 mJ/cm^2 for three times for Stratalinker).
4. We usually put ~6 million HeLa cells or ~150 mg mouse brain cortex in one microfuge tube, which will be used as one sample for the subsequent steps. It might be necessary to increase the number of cells for crosslinking if primary cells are used.
5. For each microfuge tube of crosslinked tissue (~150 mg adult mouse brain cortex or 6 ~ 10 million cultured cells), use 200 ~ 300 μL of Dynabeads Protein G. The minimum number of groups is two (one high RNase and one low RNase). If the volume of beads exceeds the maximum capacity of a microfuge tube, evenly split the beads into two or more tubes.
6. For each group, use 30 ~ 40 μg 10E10 anti-PABP for immunoprecipitation. Resuspend the beads in the original slurry volume (200 ~ 300 μL) of Antibody Binding Buffer containing anti-PABP.

7. The ideal volume of 1× TS Buffer is at least 5× tissue volume. As the usual tissue volume is <200 mg, 1 mL is usually chosen for convenience.
8. The incubator/shaker we use is an Eppendorf Thermomixer. The shaking speed might need to be optimized if a different incubator/shaker is used.
9. This and all subsequent washes are performed on a magnet rack with ice-cold buffers (kept on ice during the washes) in the same volume as the IP reaction (usually 1 mL). To speed up processing, vacuum can be used to remove the wash solutions with precautions to avoid losing beads.
10. The volume is based on 300 μL of beads per reaction. It can be scaled down if fewer amounts of beads are used. For example, reduce the total volume to 40 μL if 200 μL beads are used.
11. Radioactive materials are used. Use protection gears and collect all solid/liquid waste according to institutional guidelines for this and all subsequent steps in Subheadings 3.4–3.7.
12. We usually run the 31 kDa marker (orange) to the bottom of the gel.
13. Be sure to use autorad markers because a perfect align between the film and the membrane is essential for the subsequent RNA extraction step.
14. The adequate exposure time depends on the amount of starting materials but usually is less than 4 h for cultured cells when fresh gamma ATP is used.
15. Each group needs at least 400 μL Working Proteinase K Solution. Prepare the solution in 1.5 mL microfuge tubes for easy incubation.
16. Use 50 μL of Dynabeads Protein G per group for two rounds of purification (25 μL each). Prepare one extra reaction for the minus RT group.
17. Use 225 μL Antibody Binding buffer and 25 μL 50× Denhardt's Solution for each group. If the total volume exceeds 1 mL, use 900 μL Antibody Binding buffer and 100 μL 50× Denhardt's Solution. (The working concentration of Denhardt's solution is 5×, not 1×!) Use of Denhardt's Solution is essential for minimizing nonspecific carryover.
18. The RNA pellet is usually very small and half-opaque. It may help to use a Radiation Survey Meter to track or to confirm the presence of a pellet.
19. It is important not to over-dry the pellet. We routinely use SpeedVac to dry the pellet after taking out all liquids: 45 ~ 60 s at room temperature.
20. It is important to keep all reaction components cold for this and subsequent steps before reverse transcription. We use a

metal heat block fully surrounded by ice in an ice bucket to hold the samples.

21. The RT primers are designed to be compatible with Illumina sequencers. Each indexed primer has a unique 4-nt index, which allows the individual library to be identified from a pooled sequencing run.
22. This is a good time to take a (lunch) break.
23. To bring the reaction volume above the 25 μL minimum requirement for the G-25 columns.
24. Before spinning the samples through G-25 columns, Denhardt's Solution and 2× IP Buffer can be added to the collection tube. Then the total volume will be adjusted up to 100 μL by adding nuclease-free water.
25. It is essential to include Denhardt's Solution in the indicated wash buffers. Make a master mix for all tubes. For example, add 500 μL 50× Denhardt's Solution to 5 mL 1× IP Buffer to make the 1× IP Buffer plus 5× Denhardt's Solution for five tubes (the dilution of 50× Denhardt's Solution doesn't need to be exact).
26. Use manual pipetting to remove the wash buffer. Do not use vaccum, which will result in significant loss of beads!
27. A minimum of two rounds of cDNA immunoprecipitation is necessary to achieve the desired purity.
28. It is important to keep all reaction components cold for this and subsequent steps before PCR amplification. We use a metal heat block fully surrounded by ice in an ice bucket to hold the samples.
29. We use 8-strip PCR tubes and optically clear 8-cap strip from Bio-Rad that are compatible with real-time PCR. We cut the connection between tubes and caps with a scalpel.
30. Real-time PCR instead of conventional PCR is used for library amplification because the minimal number of cycles required for amplification has a wide range and needs to be experimentally determined. We use Bio-Rad CFX96™ Touch Real-Time PCR Detection System, which allows real-time tracking of multiple samples simultaneously and permits interruption and sample retrieval during PCR amplification.

Acknowledgments

H-W.H. is an HHMI Fellow of the Life Sciences Research Foundation. R.B.D. is an Investigator of the Howard Hughes Medical Institute. This work was also in part supported by grants from the National Institutes of Health (NS034389 and NS081706) and Simons Foundation (SFARI 240432) to R.B.D.

References

1. Tian B, Manley JL (2013) Alternative cleavage and polyadenylation: the long and short of it. Trends Biochem Sci 38:312–320
2. Derti A, Garrett-Engele P, MacIsaac KD et al (2012) A quantitative atlas of polyadenylation in five mammals. Genome Res 22:1173–1183
3. Shi Y (2012) Alternative polyadenylation: new insights from global analyses. RNA 18: 2105–2117
4. Hoque M, Ji Z, Zheng D et al (2013) Analysis of alternative cleavage and polyadenylation by 3′ region extraction and deep sequencing. Nat Methods 10:133–139
5. Di Giammartino DC, Nishida K, Manley JL (2011) Mechanisms and consequences of alter native Polyadenylation. Mol Cell 43:853–866
6. Licatalosi DD, Mele A, Fak JJ et al (2008) HITS-CLIP yields genome-wide insights into brain alternative RNA processing. Nature 456: 464–469
7. Kahvejian A, Roy G, Sonenberg N (2001) The mRNA closed-loop model: the function of PABP and PABP-interacting proteins in mRNA translation.
8. Hwang H-W, Park CY, Goodarzi H et al (2016) PAPERCLIP identifies MicroRNA targets and a role of CstF64/64tau in promoting non-canonical poly(a) site usage. Cell Rep 15: 423–435
9. He M, Liu Y, Wang X et al (2012) Cell-type-based analysis of microRNA profiles in the mouse brain. Neuron 73:35–48

Chapter 8

Microinjection and Fluorescence In Situ Hybridization Assay for Studying mRNA Export in Mammalian Cells

Ke Wang, Min Shi, and Hong Cheng

Abstract

Microinjection and Fluorescence in situ Hybridization (FISH) assay is a useful method for mRNA export studies, which can overcome the problems of traditional transfection in cells. Here, we describe the method of microinjection and FISH assay applied in investigation of mRNA export. By this method we can estimate the mRNA export kinetics, examining mRNA export in cells with low transfection efficiencies, and observing nuclear export of aberrant RNAs.

Key words Microinjection, Fluorescence in situ hybridization, mRNA export

1 Introduction

Eukaryotic mRNA export is a highly regulated step of gene expression, which is physically and functionally coupled to multi-steps of gene expression. This coupling on one hand ensures the efficiency of mRNA export, and on the other hand prevents export of improperly processed mRNAs to the cytoplasm [1–3]. DNA transfection could be used to introduce DNA constructs into the cells. However, in many cases, introducing DNA constructs into the cells by microinjection has more parent advantage for mRNA export studies. First, it could ensure the similar amount of DNA construct get into each cell in a short range of time to estimate the mRNA export kinetics (Figs. 1 and 2). Second, it could drive the DNA constructs into the cells that have low transfection efficiencies. In many cases, we found that downregulation of many mRNA export factors results in apparently low transfection efficiency. Microinjection is a suitable solution for introducing DNA into these cells (Fig. 3). Third, in order to investigate the role of certain RNA features (i.e., the 5′ cap or the 3′ polyA tail) in mRNA export,

Yongsheng Shi (ed.), *mRNA Processing: Methods and Protocols*, Methods in Molecular Biology, vol. 1648, DOI 10.1007/978-1-4939-7204-3_8, © Springer Science+Business Media LLC 2017

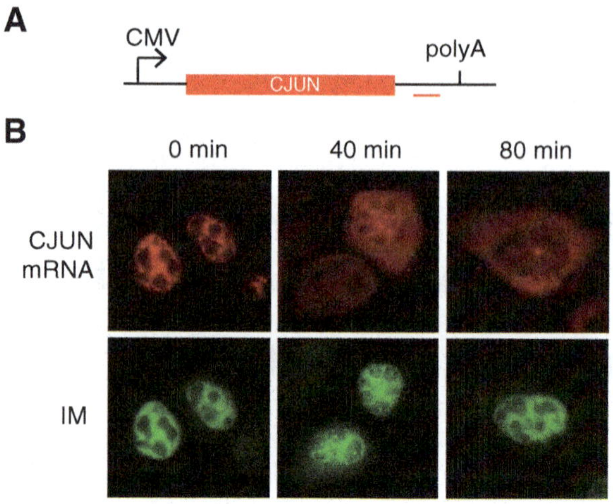

Fig. 1 The time course of CJUN mRNA export. (**a**) Schematic of the CJUN DNA construct. The CMV promoter and polyA signal site are indicated. The *red line* indicates the location of the FISH probe. (**b**) The representative cells showing the distribution of CJUN mRNA. CJUN DNA construct and injection maker (IM, *green*) were injected into the nuclei of HeLa cells. 15 min after microinjection, α-amanitin was added to block transcription and FISH was carried out with BGH probe (*red*) by after 0 min, 40 min, or 80 min the addition of α-amanitin

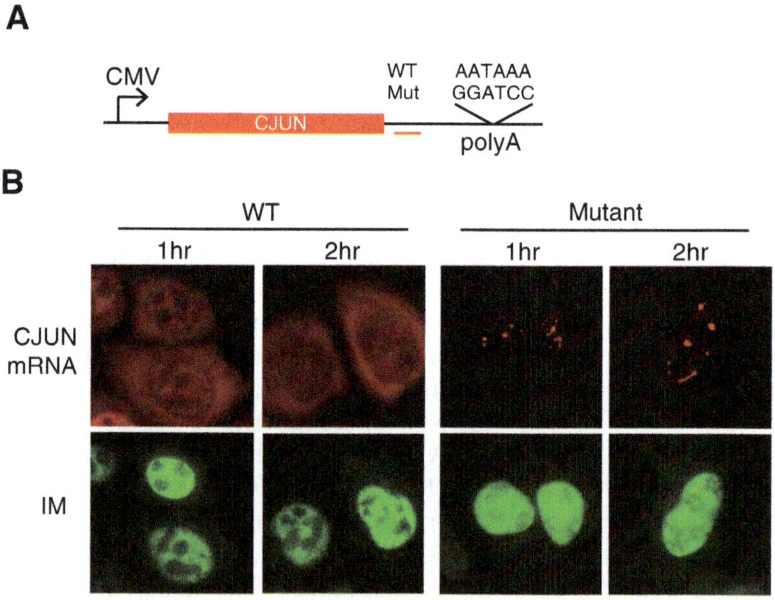

Fig. 2 PolyA signal plays a critical role in naturally intronless mRNA export. (**a**) Schematic illustration of the WT and mutant CJUN DNA constructs used in (**b**). (**b**) Equal amount of the WT or mutant CJUN DNA constructs and injection maker (IM, *green*) were injected into the nuclei of HeLa cells. 15 min after microinjection, α-amanitin was added to block transcription. FISH was carried out with BGH probe (*red*) by 1 h or 2 h after addition of α-amanitin

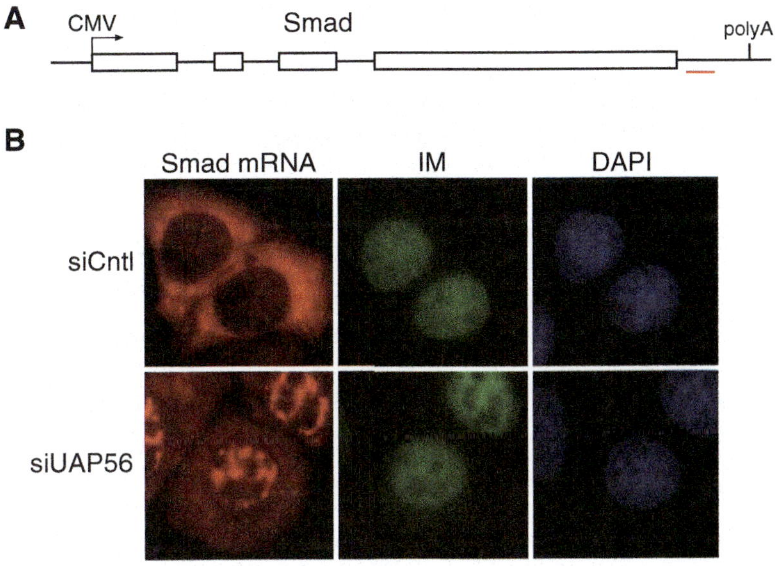

Fig. 3 Knockdown of mRNA export factor UAP56 leads to inhibition of mRNA export. (**a**) Schematic of the Smad DNA construct. The CMV promoter and polyA signal site are indicated. The *red line* indicates the location of the FISH probe. (**b**) Equal amount of Smad DNA construct and injection maker (IM, *green*) were microinjected into the nuclei of control or UAP56/URH49 knockdown cells. 15 min after microinjection, α-amanitin was added to block transcription. FISH was carried out with BGH probe (*red*) by 1 h after addition of α-amanitin

the RNAs lacking the feature could be transcribed in vitro and directly introduced into the nuclei of cells. In this case, we definitely need a way to directly get the RNAs into the nucleus. Microinjection assay followed by Fluorescence in situ hybridization (FISH) has been widely used for studying mRNA export in mammalian cells. Reporter DNA constructs that contain a eukaryotic promoter and polyadenylation signal sequence, or in vitro transcribed mRNAs, are microinjected into the nuclei of hundreds of mammalian cells in a short range of time (i.e., ~15 min). The reporter mRNAs are allowed to be transcribed (only for DNA microinjection) and processed in the nucleus. 15 min post-microinjection, the cells are treated with a transcription inhibitor (i.e., α-amanitin). Finally, at certain time points after transcription inhibition, the cells are fixed and nucleocytoplasmic distribution of the report mRNAs can be examined by FISH.

Using this system, how mRNA export is influenced by pre-mRNA processing, cis-acting export elements and the expression of mRNA export factors has been investigated [4–11].

2 Materials

2.1 Cell Preparation

1. 35 mm microinjection dishes with 20 mm coverslip at bottom.
2. DMEM with 10% Fetal Bovine Serum (FBS).

2.2 DNA Constructs and RNAs

1. DNA purification Kit.
2. mMESSAGE mMACHINE® Kit (Ambion).
3. Poly(A) Tailing Kit (Ambion).
4. MEGAclear™ Kit (Ambion).

2.3 Needles for Microinjection

1. 1.0 mm borosilicate glass capillary tubes.
2. Modle-p97 Flaming/Brown Micropipette Puller (Sutter Instrument Co.).

2.4 Microscope and Micromanipulator

1. An inverted aberration-corrected microscope with 10× eyepiece and 40× objective lens.
2. InjectMan NI-2 (Eppendrof).

2.5 Buffers and FISH Probes

2.5.1 FISH Probe

Oligonucleotides longer than 20 nucleotides and labeled with Alexa546 fluorophore are used.

2.5.2 Reagents and Solutions

Reagent	Concentration
DAPI	10 mg/mL in H_2O
Injection maker (IM)	10 mg/mL in H_2O
α-Amanitin	1 mg/mL in H_2O
Potassium acetate	3 M in H_2O

2.5.3 10× Injection Buffer (IB)

Reagent	Final concentration
Potassium chloride	1 M
HEPES	100 mM
Adjust the pH to 7.4	

2.5.4 1× PBS Buffer

Reagent	Final concentration
Sodium chloride	137 mM
Potassium chloride	2.7 mM
Sodium phosphate dibasic	10 mM
Monopotassium phosphate	1.8 mM
Adjust the pH to 7.4	

2.5.5 SSC-Formamide Buffer

Reagent	Final concentration
Sodium chloride	150 mM
Trisodium citrate	15 mM
Formamide	50%
Adjust the pH to 7.0	

2.5.6 Hybridization Buffer (HB)

Reagent	Final concentration
Sodium chloride	150 mM
Trisodium citrate	15 mM
Formamide	50%
Dextran sulfate sodium salt	100 mg/mL
Bovine serum albumin	20 ng/mL
Yeast tRNA	1 mg/mL
Vanadyl ribonucleoside complexes	3.66 mg/mL
Adjust the pH to 7.0	

(*see* **Note 1**).

3 Methods

3.1 HeLa Cell Preparation

1. 1.2×10^5 of HeLa cells are seeded on the glass bottom.
2. The cells are incubated at 37 °C with 5% CO_2 for more than 12 h while the cells are sprawled completely.
3. Before microinjection, the cells should be covered by fresh DMEM medium containing 10% FBS.

3.2 Preparation of Tips and Needles for Microinjection

3.2.1 Preparation of Tips

1. Heat the 200 μL pipette tips as the plastics being molten.
2. Stretch the tip of pipette tips to a thin pipe that could drive into the needle.
3. Store the tips at room temperature as the thin pipe up in a clean place for use later.

3.2.2 Preparation of Needles

1. Set up the arguments of micropipette puller according to the manufacturer's protocol (*see* **Notes 2** and **3**).
2. Store the needles at the room temperature in a clean place for use later.

3.3 Plasmid Microinjection

1. The plasmid for microinjection should be pure (*see* **Note 4**) and contains the promotor that could be transcribed in mammalian

cells (e.g., CMV promotor). Prepare the plasmid using the DNA purification Kit and elute the plasmid with H_2O. Measure the concentration and quality of the plasmid with Nanodrop. Store the plasmid at −80 °C for long time use.

2. Mix the plasmid and 10× Injection buffer in H_2O with or without injection marker (FITC-conjugated 70 kDa dextran). The final concentration of the plasmid is 100 ng/μL. The mixture should be centrifuged at 145,000 × *g* for 40 min before microinjection.
3. Pipette 3 μL of the mixture from the top of the liquid without disturbing the bottom with a tip prepared in Subheading 3.2.1.
4. Load the mixture into a needle prepared in Subheading 3.2.2 and fix the needle on arm of InjectMan NI-2.
5. Find the tip of the needle in the horizon of eyepiece and move the tip of the needle to the middle of the horizon.
6. Take a brief break of the needle so that the liquid could brow out.
7. Set the arguments of InjectMan NI-2 and fix the location of the needle where the needle could pierce the nucleus of the HeLa cells just deep enough.
8. Optimize the arguments of InjectMan NI-2, finally the nucleus should be slightly dilated with once injection.
9. Inject all cells in the horizon from one to another. Finally, the total cells should be more than 300 cells.
10. Incubate the dishes at 37 °C degree with 5% CO_2 for desired period of time before fixation (*see* **Note 5**).

3.4 Microinjection of PCR Products

1. The PCR products containing the whole CMV promotor and polyA signal for microinjection should be pure (*see* **Note 6**).
2. The PCR products are amplified with high-fidelity DNA polymerase.
3. The PCR products are collected by the agarose gel electrophoresis and gel extraction. The PCR products should be eluted by H_2O to a final concentration more than 250 ng/μL. The PCR products are stored at −80 °C for long time use.
4. The mixture of PCR products for microinjection is prepared as described in Subheading 3.3, **step 2** with slight modification. The final concentration of PCR products in the mixture should be 200 ng/μL.
5. The part of microinjection is described as Subheading 3.3.

3.5 RNA Microinjection

1. The mature mRNAs for microinjection are synthesized in vitro using mMESSAGE mMACHINE® Kit, the polyA tail were adding with Poly(A) Tailing Kit using mMESSAGE mMACHINE® Kit reaction products.

2. The mRNAs are purified with MEGAclear™ Kit and precipitation of mRNAs with 3 M KAc and ethanol (*see* **Note 7**). The mRNAs are dissolved in H_2O with the final concentration more than 1500 ng/μL. The mRNAs are stored at −80 °C for long time use.
3. The mixture of mRNAs for microinjection is prepared as described in Subheading 3.3, **step 2** with some modification. The final concentration of mRNAs in the mixture should be 1000 ng/μL. The Injection marker must be included in the mixture.
4. The part of microinjection is described as Subheading 3.3. The operation of microinjection should be very careful ensuring the mixture is injected exactly in the nucleus without any leakage.

3.6 FISH

1. Fix the cells on the coverslip at bottom of dishes with 4% PFA in 1× PBS for 15 min at the room temperature.
2. Permeabilize the cells with 1× PBS/0.1% Triton for 15 min at the room temperature.
3. Wash the cells twice with 1× SSC/50% Formamide following three washes with 1× PBS.
4. The cells are incubated with 70 nt DNA probe (0.75 ng/μL), which specifically hybridize to some vector sequence, at 37 °C for 12 h (*see* **Notes 8** and **9**).
5. Wash the cells four times with 1× SSC/50% Formamide. Each time the cells should stay in 1× SSC/50% Formamide for 10 min at the room temperature.
6. Stain the nucleus with DAPI (0.01 ng/μL) in 1× PBS for 2 min at the room temperature.
7. Wash the cells three times with 1× PBS.
8. Images are captured with CCD camera on an inverted microscope and FISH quantitation is carried out using ImageJ software.

4 Notes

1. HB stored at 4 °C can last for less than 2 months.
2. In micropipette puller, the Pt ring is the critical part. While pulling the needles, avoid touching the ring to keep the arguments consistent.
3. The fixation of quartz tube to the micropipette puller should be kept the same tightness.
4. The plasmid for microinjection should be eluted from the membrane in Kit with H_2O directly. Precipitation of plasmids with 3 M NaAc and ethanol is forbidden.

5. To monitor the mRNA export efficiency, transcription should be blocked with α-amanitin (4 ng/μL) in 10% FBS. α-amanitin treatment must be less than 4 hr. to avoid toxicity.
6. The PCR products for microinjection are dissolved in H_2O. The agarose could not be present in the solution. Precipitation of PCR products with 3 M NaAc and ethanol is forbidden.
7. For RNA injection, mRNAs must be precipitated with 3 M KAc and ethanol.
8. The DNA probe is labeled at the 5′ end with Alexa Fluor 546 NHS Ester and high-performance liquid chromatography purified. It is dissolved in H_2O to a final concentration of 100 ng/μL. The probe should be aliquoted and stored at −20 °C avoiding light.
9. The dishes should be placed in a humid chamber at 37 °C avoiding HBS drying that might result in high background for FISH.

References

1. Maniatis T, Reed R (2002) An extensive network of coupling among gene expression machines. Nature 416(6880):499–506. doi:10.1038/416499a
2. Pandit S, Wang D, XD F (2008) Functional integration of transcriptional and RNA processing machineries. Curr Opin Cell Biol 20(3):260–265. doi:10.1016/j.ceb.2008.03.001
3. Reed R (2003) Coupling transcription, splicing and mRNA export. Curr Opin Cell Biol 15(3):326–331
4. Lei H, Dias AP, Reed R (2011) Export and stability of naturally intronless mRNAs require specific coding region sequences and the TREX mRNA export complex. Proc Natl Acad Sci U S A 108(44):17985–17990. doi:10.1073/pnas.1113076108
5. Palazzo AF, Springer M, Shibata Y, Lee CS, Dias AP, Rapoport TA (2007) The signal sequence coding region promotes nuclear export of mRNA. PLoS Biol 5(12):e322. doi:10.1371/journal.pbio.0050322
6. Dias AP, Dufu K, Lei H, Reed R (2010) A role for TREX components in the release of spliced mRNA from nuclear speckle domains. Nat Commun 1:97. doi:10.1038/ncomms1103
7. Lei H, Zhai B, Yin S, Gygi S, Reed R (2013) Evidence that a consensus element found in naturally intronless mRNAs promotes mRNA export. Nucleic Acids Res 41(4):2517–2525. doi:10.1093/nar/gks1314
8. Valencia P, Dias AP, Reed R (2008) Splicing promotes rapid and efficient mRNA export in mammalian cells. Proc Natl Acad Sci U S A 105(9):3386–3391. doi:10.1073/pnas.0800250105
9. Chi B, Wang Q, Wu G, Tan M, Wang L, Shi M, Chang X, Cheng H (2013) Aly and THO are required for assembly of the human TREX complex and association of TREX components with the spliced mRNA. Nucleic Acids Res 41(2):1294–1306. doi:10.1093/nar/gks1188
10. Chi B, Wang K, Du Y, Gui B, Chang X, Wang L, Fan J, Chen S, Wu X, Li G, Cheng H (2014) A sub-element in PRE enhances nuclear export of intronless mRNAs by recruiting the TREX complex via ZC3H18. Nucleic Acids Res 42(11):7305–7318. doi:10.1093/nar/gku350
11. Akef A, Zhang H, Masuda S, Palazzo AF (2013) Trafficking of mRNAs containing ALREX-promoting elements through nuclear speckles. Nucleus 4(4):326–340. doi:10.4161/nucl.26052

Chapter 9

Methods for the Detection of Adenosine-to-Inosine Editing Events in Cellular RNA

Eimile Oakes, Pranathi Vadlamani, and Heather A. Hundley

Abstract

Modification of RNA is essential for properly expressing the repertoire of RNA transcripts necessary for both cell type and developmental specific functions. RNA modifications serve to dynamically re-wire and fine-tune the genetic information carried by an invariable genome. One important type of RNA modification is RNA editing and the most common and well-studied type of RNA editing is the hydrolytic deamination of adenosine to inosine. Inosine is a biological mimic of guanosine; therefore, when RNA is reverse transcribed, inosine is recognized as guanosine by the reverse transcriptase and a cytidine is incorporated into the complementary DNA (cDNA) strand. During PCR amplification, guanosines pair with the newly incorporated cytidines. As a result, the adenosine-to-inosine (A-to-I) editing events are recognized as adenosine to guanosine changes when comparing the sequences of the genomic DNA to the cDNA. This chapter describes the methods for extracting endogenous RNA for subsequent analyses of A-to-I RNA editing using reverse transcriptase-based approaches. We discuss techniques for the detection of A-to-I RNA editing events in messenger RNA (mRNA), including analyzing editing levels at specific adenosines within the total pool of mRNA versus analyzing editing patterns that occur in individual transcripts and a method for detecting editing events across the entire transcriptome. The detection of RNA editing events and editing levels can be used to better understand normal biological processes and disease states.

Key words RNA editing, A-to-I editing, ADARs, Inosine, RNA modifications, Transcriptome-wide sequencing, High-throughput sequencing

1 Introduction

RNA editing is a process in which single nucleotides within mRNA sequences are selectively modified by specialized enzymes [1, 2]. The most common type of RNA editing in animals is adenosine-to-inosine (A-to-I) RNA editing, which is catalyzed by adenosine deaminases that act on RNA (ADARs) (Fig. 1) [3, 4]. ADARs edit adenosines within double-stranded regions of RNA, including viral RNAs [5], small regulatory RNAs [6], and double-stranded regions of mRNAs [7, 8], the latter of which will be the focus of the methods described in this chapter. The double-stranded RNA (dsRNA) substrates in mRNA can be located in coding regions and

Yongsheng Shi (ed.), *mRNA Processing: Methods and Protocols*, Methods in Molecular Biology, vol. 1648,
DOI 10.1007/978-1-4939-7204-3_9, © Springer Science+Business Media LLC 2017

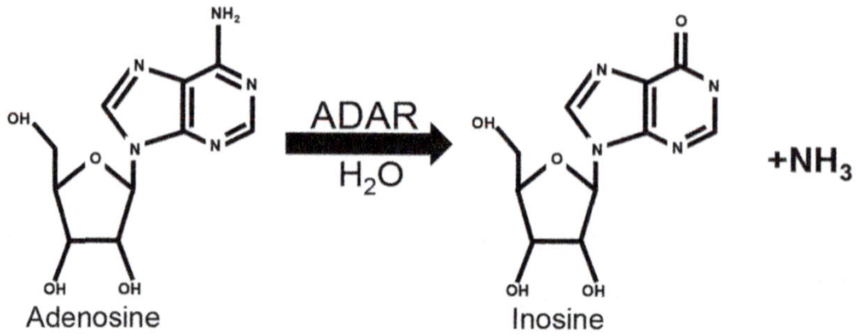

Fig. 1 Adenosine-to-inosine RNA editing. ADARs catalyze the hydrolytic deamination of the C6 position of adenosine creating inosine. This deamination occurs within double-stranded RNA

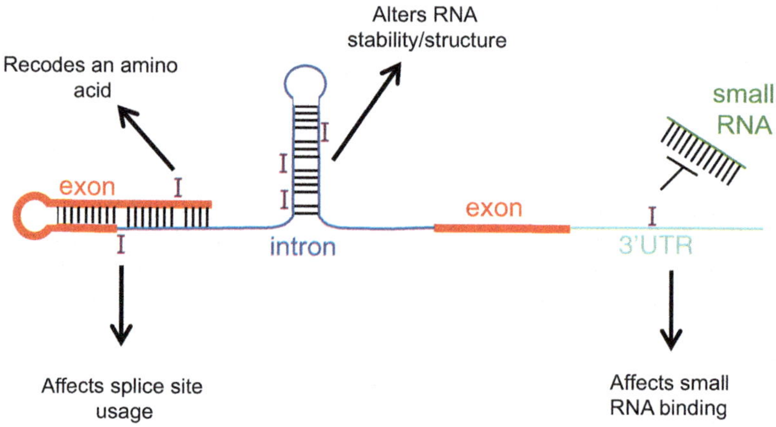

Fig. 2 Targets and outcomes of A-to-I RNA editing. ADARs target double-stranded regions within pre-mRNA. Depending on the region of dsRNA targeted by an ADAR enzyme, different outcomes can occur. For example, editing within an exon can lead to recoding of an amino acid; editing at a splice donor site can affect splice site selection; editing within introns and UTRs can affect RNA stability and structure; and editing at a small RNA binding site can affect binding of small RNAs

formed by base-pairing between introns and exons or long dsRNA structures contained completely within the noncoding regions of the mRNA, including introns and untranslated regions (UTRs) (Fig. 2). A-to-I editing modifications can regulate both protein function and expression through the recoding of amino acid sequences, altering of splice site usage, affecting small RNA binding sites, or altering RNA stability/structure (Fig. 2).

A-to-I RNA editing is widespread in mRNAs, with over two million editing events occurring within the human transcriptome [9–13]. Furthermore, aberrant editing is common in many human diseases including cancers and neuropathological diseases [14, 15]. Thus, the detection of RNA editing events and changes in editing levels is critical for understanding both normal biological processes and disease states.

Single-nucleotide changes in RNA transcripts can be detected by comparing the sequence of endogenous RNA that has been isolated and reverse transcribed to either the sequence of genomic DNA or RNA from an editing enzyme-deficient system. Depending upon the A-to-I editing sites of interest, including the number of events to analyze, the novelty of the sites and the co-occurrence of editing sites on an individual transcript, the method to detect editing sites can include Sanger sequencing, high-throughput sequencing of specific transcripts and/or transcriptome-wide RNA-seq.

The most commonly used technique to detect RNA editing is reverse transcriptase-polymerse chain reaction (RT-PCR) followed by Sanger sequencing. Inosine is a biological mimic of guanosine. When the reverse transcriptase encounters inosine, cytidine will be incorporated into the reverse strand. Upon amplification, these cytidine residues will base-pair with guanosine. Thus, when the cDNA is amplified, all A-to-I editing events can be visualized as A-to-G changes in the amplified cDNA. The basic strategy is to isolate RNA and reverse transcribe the RNA into cDNA (Fig. 3a). Depending upon the number of genes to analyze, structure of the RNA of interest and expression level of the RNA of interest,

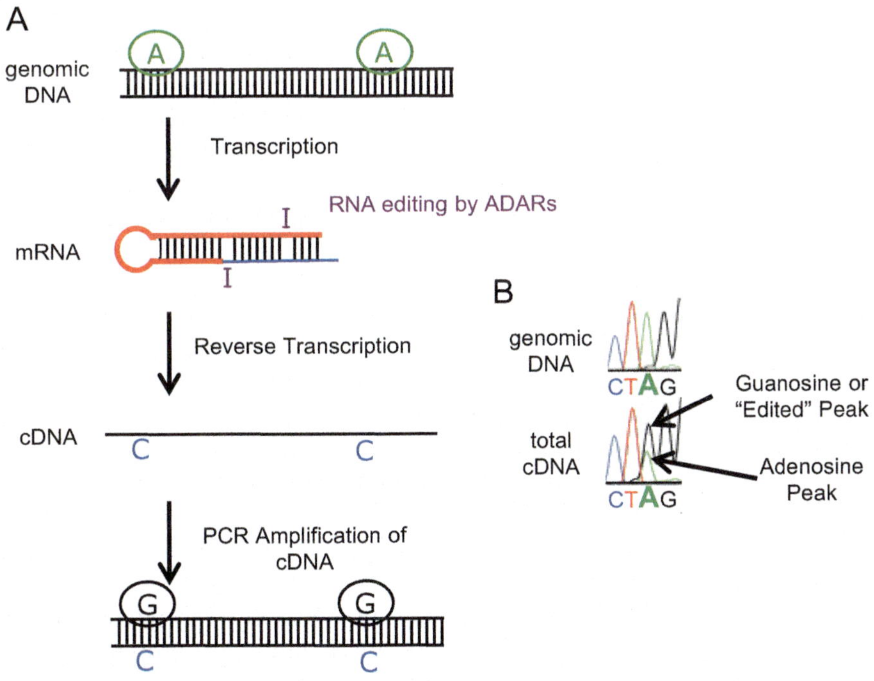

Fig. 3 Identifying A-to-I editing events. (**a**) RNA editing can be measured by comparing the genomic DNA sequence to the amplified cDNA sequence. RNA editing occurs on transcribed mRNAs. To determine editing, RNA must be isolated, reverse transcribed into cDNA, amplified by PCR, and Sanger sequenced. (**b**) After Sanger sequencing, the chromatagrams from sequenced genomic DNA and cDNA can be compared. The editing level can be determined by measuring the height of the guanosine peak and dividing it by the height of the guanosine peak plus the adenosine peak

the reverse transcription conditions can vary. A gene-specific primer is ideal for reverse transcribing transcripts that are lowly expressed or have long dsRNA structures that are more difficult to reverse transcribe. Random hexamers work fine for more highly expressed genes and are beneficial because amplification of multiple genes can be done using the same cDNA and reverse transcription reaction.

As there are currently no methods to sequence RNA, the cDNA must be amplified to double-stranded DNA. To analyze a specific editing event, primers must be designed to amplify the region of cDNA that contains the editing site. For example, analysis of the editing level of an adenosine within the coding region of a mature mRNA would require the use of primers that span the exons surrounding the site and not include any intronic sequence. In order to accurately determine editing levels, it is essential to also PCR amplify a negative control sample of RNA that has not been reverse transcribed, to ensure that there is no contaminating genomic DNA in the starting RNA, and thus the PCR amplification is only of the newly synthesized cDNA. After PCR amplification, the cDNA can be sequenced using next generation (discussed in more detail in the following paragraph) or Sanger sequencing. For Sanger sequencing, comparing the resulting sequence chromatograms of the amplified cDNA to the genomic DNA sequence can identify editing events. Sanger sequencing chromatograms are a series of colored peaks that indicate the presence of a specific nucleotide at each position within the cDNA that was amplified (Fig. 3b). Generally, a single peak of one color is present at each location within the genomic DNA and the same nucleotide is seen in the chromatogram of the amplified cDNA. However, at sites that are modified by editing enzymes, peaks of two different colors—one color representing an adenosine peak in the genomic DNA and a second color representing the guanosine peak in the amplified cDNA—will be observed. However, it is important to note that the majority of A-to-I editing events occur only on a fraction of cellular transcripts; therefore, the Sanger sequencing chromatogram of a given transcript may contain two peaks at the same site. The relative height of these peaks to each other can be measured to determine the percentage of transcripts containing each of the nucleotides and thus, the editing level at that specific adenosine.

One interesting and novel method to detect RNA editing events on specific transcripts is high-throughput sequencing of a specific transcript. Unlike Sanger sequencing a pool of RNA transcripts, the RT-PCR and high-throughput sequencing method allows for the detection of editing events and editing patterns that occur on a single transcript [16–18]. Another method for identifying editing events is through transcriptome-wide sequencing of reverse-transcribed RNA. The primary benefit of this method is that editing events in thousands of transcripts can be identified rather than a single or a handful of transcripts. In addition, this method is the most robust for identifying novel editing sites. However, developing a stringent bioinformatics pipeline for the accurate detection of editing events is essential [19]. Furthermore, several editing events identified by the pipeline must be verified by Sanger sequencing in order to determine the false-discovery rate.

2 Materials

All chemicals utilized should be of molecular biology grade (DNase-, RNase-, and Protease-free) and solutions should be prepared in ultrapure water. For RNA isolation and reverse transcription reactions, perform all steps in an RNase-free environment by first cleaning spaces with 95% ethanol and RNase removal solution (such as RNase Zap (Sigma)), wearing gloves at all times and using disposable sterile pipettes, barrier tips for pipetteman, glassware that has been baked at 200°C for at least 24 h, and reagents/chemicals that are reserved for RNA use (*see* **Note 1**) from the time they are first opened.

All of the methods described in this chapter utilize the property that reverse transcriptase recognizes inosine as guanosine and, thus inserts a cytidine into the cDNA product at each editing site. However, cytidine insertions at individual sites in the RNA or cDNA could also be a result of errors in RNA polymerase II and reverse transcriptase, respectively. Additionally, as PCR amplification and sequencing is utilized in all of these detection methods, these single-nucleotide changes can be a result of errors in DNA polymerase or the sequencing reaction. Therefore, to accurately conclude that a given adenosine undergoes A-to-I RNA editing, the experimental design should include isolation of RNA from an editing deficient or editing reduced control. In some model organisms, such as *C. elegans* or *D. melanogaster*, ADARs are not essential and thus RNA isolated from knockout organisms can be used alongside the condition of interest. In mammals, where ADARs are essential, RNAi knockdown cell lines can be used for a condition with reduced A-to-I editing.

2.1 Isolation of RNA and Genomic DNA

2.1.1 RNA Isolation

1. TRIzol Reagent.
2. 70% Ethanol.
3. Isopropanol.
4. Chloroform.
5. TURBO DNase, supplied with 10× TURBO DNase buffer.
6. RNasin (RNase inhibitor).
7. QIAGEN RNeasy kit: QIAGEN.

2.1.2 DNA Isolation

1. Ethanol, 100% and 75%.
2. 0.1 M Sodium Citrate in 10% Ethanol, pH 8.5.
3. 8 mM NaOH.
4. 1 M HEPES, pH 7.5.

2.2 Identifying RNA Editing Sites by RT-PCR and Sanger Sequencing

2.2.1 Reverse Transcription

1. Reverse transcription primer: 5 μM in RNA water (*see* **Note 2**).
2. dNTPs: 10 mM in RNA water.
3. ThermoScript Reverse Transcriptase: Invitrogen. Supplied with 5× cDNA Synthesis Buffer and 0.1 M DTT.
4. RNasin (RNase Inhibitor).
5. RNase H.

2.2.2 Polymerase Chain Reaction

1. 1st PCR 5′ primer: 5 μM in water.
2. 1st PCR 3′ primer: 5 μM in water.
3. 2nd PCR 5′ primer: 5 μM in water.
4. 2nd PCR 3′ primer: 5 μM in water.
5. 10 mM dNTPs.
6. Platinum *Pfx* DNA Polymerase: Invitrogen. Supplied with 10× *Pfx* Buffer, 50 mM $MgSO_4$, and PCR Enhancer solution.
7. QIAquick Gel Extraction Kit: QIAGEN.

2.2.3 Sanger Sequencing

1. Sequencing primer: 5 μM in water.
2. 5 mM $MgCl_2$.
3. BigDye Terminator v3.1: Applied Biosystems.

2.3 Identifying Editing Sites by RT-PCR and High-Throughput Sequencing

2.3.1 Reverse Transcription and RT-PCR

1. All materials described in Subheadings 2.2.1 and 2.2.2, with the exception of the QIAquick Gel Extraction Kit.
2. QIAquick PCR Purification Kit: QIAGEN.

2.3.2 Library Preparation

1. T4 DNA Polymerase.
2. Klenow DNA Polymerase.
3. T4 Polynucleotide Kinase.
4. 10× T4 DNA Ligase buffer with 10 mM ATP.
5. 10 mM dNTPs.
6. QIAquick PCR Purification Kit with elution buffer: QIAGEN.
7. Klenow Fragment (3′-5′ exonuclease minus).
8. 10× Klenow Buffer (*see* **Note 3**).

9. 5 mM dATP.
10. MinElute PCR Purification Kit with elution Buffer: QIAGEN.
11. T4 DNA ligase.
12. Tru-Seq adapters: Illumina.
13. AMPure XP magnetic beads: Agencourt.
14. Magnetic eppendorf stand, such as DynaMag2: Life Technologies.
15. 70% ethanol, stored at −20°C.

2.4 Identifying RNA Editing Sites by Transcriptome-Wide RNA Sequencing

2.4.1 Enrichment of Polyadenylated mRNA Using Magnetic Oligo (dT) Beads

1. Dynabeads oligo (dT): Invitrogen.
2. Magnetic eppendorf stand, such as DynaMag2: Life Technologies.
3. Binding buffer: 20 mM Tris–HCl pH 7.5, 1.0 M LiCl, 2 mM EDTA.
4. Wash buffer: 10 mM Tris–HCl pH 7.5, 0.15 M LiCl, 1 mM EDTA.
5. 10 mM Tris–HCl pH 7.8.

3 Methods

3.1 Isolation of RNA and Genomic DNA

High-quality and DNA-free RNA is essential for accurately measuring RNA editing levels. Genomic DNA cannot be edited; therefore, DNA contamination will add to the adenosine signal, thus skewing the assessment of the RNA editing level. Furthermore, isolation of genomic DNA is important because genomic DNA can be used as a control in the assays to ensure that the A-to-G changes in the sequenced cDNA are the result of RNA editing, rather than single-nucleotide polymorphisms or other genomic alterations.

3.1.1 RNA Isolation

1. Harvest the cells or tissue in 1 mL of TRIzol reagent (*see* **Note 4**) and freeze the mixture in liquid nitrogen. To isolate the RNA, thaw the mixture in a 37°C water bath and then vortex for 1 min. Repeat the freeze-thaw cycle twice.
2. Add 200 μL of chloroform and mix by inverting the tube 15 times by hand.
3. Incubate the mixture at room temperature for 5 min, followed by centrifugation in a table-top centrifuge at 3500 × *g* for 15 min at 4°C.
4. After centrifugation, the mixture will separate into three phases with a clear layer containing the RNA on top, a thin white layer containing the DNA, and a red phenol phase containing the proteins and lipids. Carefully remove the clear RNA phase and transfer to a new tube without disturbing the other phases (save the remaining phases for DNA isolation) (*see* **Note 5**).

5. Add an equal volume of isopropanol to the RNA phase and incubate at room temperature for 10 min.
6. Centrifuge the mixture in a microcentrifuge at >15,000 × *g* at 4°C for at least 30 min to precipitate the RNA.
7. After centrifugation, remove the liquid without disturbing the RNA pellet, add 1 mL of 70% ethanol, and centrifuge for 5 min at 4°C.
8. Without disturbing the RNA pellet, completely remove all the ethanol and add 85 μL of RNA water. Completely resuspend the RNA pellet in water and measure the RNA concentration using a Nanodrop spectrometer.
9. To remove any trace amount of DNA contamination from the RNA preps, incubate the following reaction at 37°C for 1 h (*see* **Note 6**).

100 μg of total RNA	__ μL
10× TURBO DNase buffer	10 μL
RNasin	2 μL
TURBO DNase	5 μL
RNA water (up to 100 μL)	__ μL

10. After incubation, purify the RNA from the DNase using the QIAGEN RNeasy kit.

3.1.2 DNA Isolation

1. Using the remaining phases from Subheading 3.1.1, **step 3**, discard any of the remaining clear aqueous layer without disturbing the white DNA layer. Add 0.3 mL of 100% ethanol to remaining phases and mix by inverting five times.
2. Incubate for 3 min at room temperature and then centrifuge the mixture at 2000 × *g* for 5 min at 4°C to precipitate the DNA.
3. Without disturbing the DNA pellet, remove the phenol and ethanol from the tube and wash the pellet with 1 mL of 0.1 M Sodium Citrate in 10% ethanol, pH 8.5.
4. Incubate the mixture at room temperature for 30 min, inverting by hand every 10 min.
5. Centrifuge at 2000 × *g* for 5 min at 4°C. Repeat the wash and centrifugation steps two more times.
6. Add 2 mL of 75% ethanol and incubate for 15 min at room temperature, inverting by hand every 5 min.
7. Centrifuge at 2000 × *g* for 5 min at 4°C. Remove the liquid and air-dry the DNA pellet.
8. Resuspend the DNA pellet in 300–600 μL of 8 mM NaOH by vortexing and pipetting. Add 50 μL of 1 M HEPES, pH 7.5 per 250 μL of NaOH to bring the pH to neutral.

3.2 Identifying RNA Editing Sites by RT-PCR and Sanger Sequencing

Sanger sequencing is a simple method for identifying the level of posttranscriptional nucleotide modifications on total mRNA, including A-to-I editing events. However, mRNA cannot be directly sequenced via Sanger sequencing and must be reverse transcribed to complementary DNA (cDNA), which can be compared to the genomic DNA to determine editing levels.

3.2.1 Primer Design

Since A-to-I editing events occur in double-stranded regions of RNA and these double-stranded structures can inhibit/reduce the efficiency of reverse transcriptase, it is important to understand the RNA template before designing reverse transcription and PCR primers (*see* **Note 7**).

1. Predict the secondary dsRNA structure using the RNA Folding Form Application from the mFOLD Web Server [20] (http://unafold.rna.albany.edu/?q=mfold/RNA-Folding-Form) (Fig. 4).
2. Design a reverse transcription primer that binds outside of the double-stranded region and outside of the editing events (*see* **Note 8**).
3. Design a pair of forward and reverse primers that amplify the double-stranded region that contains the editing event(s). It is best if the primers fall outside the double-stranded region (*see* **Note 9**).

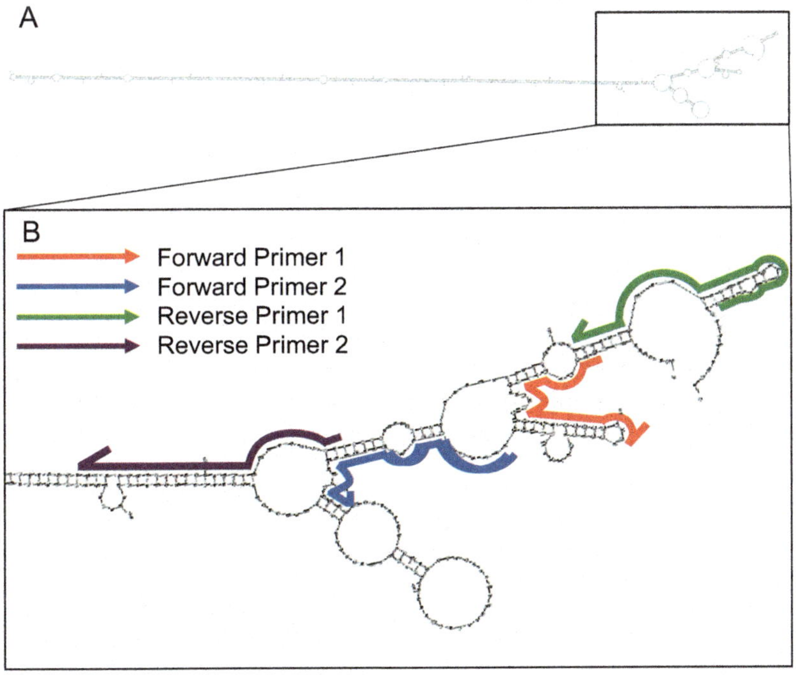

Fig. 4 Designing primers to amplify double-stranded cDNA. (**a**) mfold prediction of double-stranded structure of *C. elegans daf-2* intronic region. (**b**) Two sets of nested primers shown as colored arrows amplify the double-stranded region

4. Depending on the mRNA abundance, it may be necessary to do a second PCR amplification of the cDNA to have sufficient quantities for sequencing. Therefore, a second set of primers nested within the first set can be designed following the same guidelines as when designing the first set of primers (*see* **Note 10**).
5. Design a sequencing primer for Sanger sequencing (*see* **Note 11**).

3.2.2 PCR Annealing Temperature Optimization

1. Set up annealing temperature gradient PCR reactions with both sets of the primers designed above (*see* **Note 12**). The template for the reaction can be either genomic DNA or cDNA (*see* **Note 13**).

10× *Pfx* buffer	2.5 μL
50 mM $MgSO_4$	0.5 μL
PCR enhancer solution	3.5 μL
10 mM dNTP	1 μL
5 μM 5′ primer	2 μL
5 μM 3′ primer	2 μL
cDNA or genomic DNA	1–3 μL
Platinum *Pfx* DNA polymerase	0.5 μL
Water up to 25 μL	__ μL

PCR Program for Platinum *Pfx* DNA Polymerase

1. Denaturing step	94°C	5 min	1 cycle
2. Denaturing	94°C	15 s	
3. Annealing	45–65°C gradient	30 s	
4. Extension	68°C	15 s/250 bp	
5. Final extension step	68°C	5 min	1 cycle
6. Hold	4°C	Forever	

Repeat **steps 2–4** 29 times for a total of 30 cycles.

2. Subject products to agarose gel electrophoresis and determine the optimal temperature for PCR by selecting the annealing temperature that produces the brightest band with faintest nonspecific bands (*see* **Note 14**) (Fig. 5).

3.2.3 Reverse Transcription

1. Mix 2 μg of RNA that has been DNased (from Subheading 3.1.1) with RNA water to bring the volume up to 5 μL (*see* **Note 15**). Add the following premix:

RT primer	2 μL
10 mM dNTP	2 μL

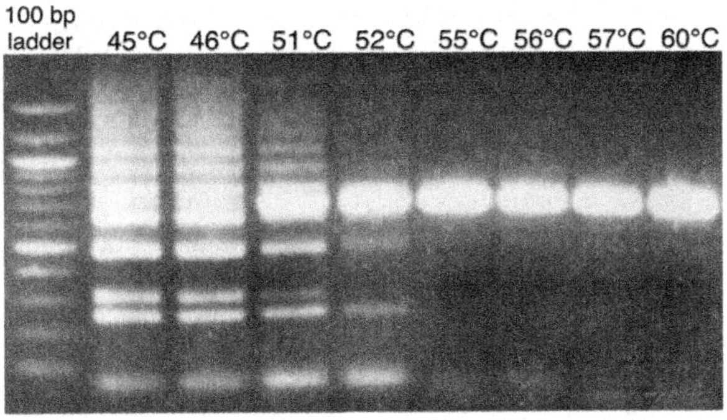

Fig. 5 Agarose gel image of PCR products of the intronic sequence of *C. elegans rab-3* amplified from genomic DNA. PCR was performed with a temperature gradient (temperatures listed above each well) to determine optimal annealing temperature of primers

2. Make mixes for the RT and minus RT samples (*see* **Note 16**).
   (a) Substitute RNA water for ThermoScript in the minus RT mix.

5× cDNA synthesis buffer	4 μL
0.1 M DTT	1 μL
RNasin	1 μL
ThermoScript RT	1 μL
RNA water	1 μL

3. Incubate the RNA/primer/dNTP mixture from **step 1** at 65°C for 5 min to disrupt the structured regions of RNA.
4. Incubate the reaction on ice for 1 min (*see* **Note 17**).
5. Add the mixes from **step 2** and incubate the reaction mixture at 52°C for 1 h to synthesize the cDNA.
6. Heat inactivate the reverse transcriptase by incubating the reaction mixture at 85°C for 10 min (*see* **Note 18**).
7. Degrade any remaining DNA-RNA hybrids by adding 1 μL of RNase H and incubating the reaction mixture at 37°C for 20 min.
8. The cDNA can be stored at −20°C.

3.2.4 Polymerase Chain Reaction

1. Using the conditions described in Subheading 3.2.2, perform the first round of PCR on cDNA (from Subheading 3.2.3), a minus RT negative control and a positive control for the PCR (1 μL of genomic DNA) with a final volume of 25 μL (*see* **Note 19**).
2. Subject 20 μL of the PCR reaction to agarose gel electrophoresis (Fig. 6). An ideal editing assay will have a single PCR product and no band in the minus RT control lane.
3. (Optional) If the first PCR does not produce bright visible bands on the agarose gel, perform a second PCR that is a total

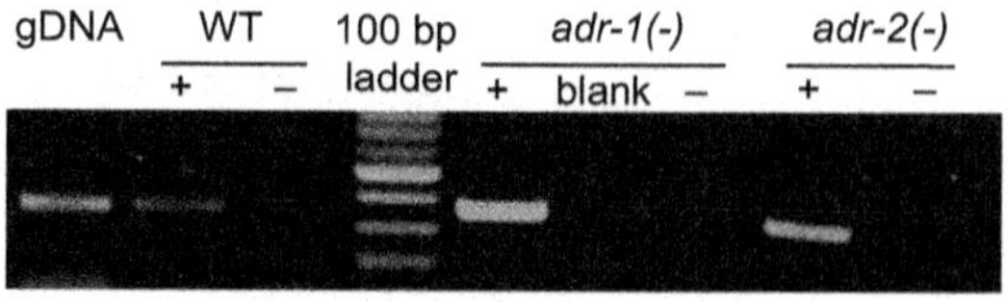

Fig. 6 Agarose gel image of PCR products from the amplification of *C. elegans lam-2* 3′ UTR. PCR products from cDNA (+) or genomic DNA (gDNA) are observed. RT-PCR performed using minus RT control reactions (–) do not show any amplified product indicating no DNA contamination in RNA samples

volume of 50 μL. For this PCR, use the second set of optimized primers, which should target sequences contained within the first PCR product (*see* blue and purple arrows in Fig. 4b) and the first PCR product as the template (*see* **Note 20**).

10× *Pfx* buffer	5 μL
50 mM $MgSO_4$	1 μL
PCR enhancer solution	7 μL
10 mM dNTP	2 μL
5 μM 5′ primer (second)	4 μL
5 μM 3′ primer (second)	4 μL
1st PCR product	0.5 μL
Platinum *Pfx* DNA polymerase	0.5 μL
Water	26 μL

4. Gel extract the PCR product using the QIAquick gel extraction kit.

3.2.5 Sanger Sequencing

1. Mix the recommended amount of gel extracted RT-PCR product (*see* **Note 21**) with water to bring the volume up to 5.5 μL. Add the following premix:

5 mM $MgCl_2$	3 μL
5 μM sequencing primer	1 μL
ABI Big Dye version 3.1	0.5 μL

2. In a PCR machine, perform the sequencing reaction below:

Sequencing Program for ABI Big Dye

1. 95°C	5 min	1 cycle
2. 95°C	30 s	
3. 50°C	10 s	
4. 60°C	4 min	
5. 4°C	Forever	

Repeat **steps 2–4** 39 times for a total of 40 cycles.

3. Subject the sequencing reaction to capillary electrophoresis. This is typically done at a sequencing core in an academic setting or at sequencing company, such as Genewiz (https://www.genewiz.com/en).

3.2.6 Detection of Editing Sites in the Sequence Chromatogram

1. Obtain the sequencing chromatogram files (.ab1) from the sequencing facility.
2. Align the sequencing chromatograms to a reference sequence using a program such as CodonCode Aligner (http://www.codoncode.com/aligner/).
3. As most A-to-I editing events occur at levels less than 100%, editing events can be identified by scanning the chromatograms manually and looking for peaks that have an overlap of both A and G at a given nucleotide (*see* **Note 22**).
4. To be identified as a true A-to-I editing event, the editing site must meet at least one of these two criteria:
   (a) The chromatogram from the genomic DNA indicates the presence of an A nucleotide at a certain position while a mixture of A and G (or G if the editing is equal to 100%) can be observed in that position in the RT-PCR product.
   (b) A mixture of A and G peaks at a given position in the RT-PCR products from wild-type RNA is seen as only an A peak at that position in the RT-PCR products from ADAR-deficient cells/tissues/animals.
5. Determine the percent editing by measuring the heights of the A and G peaks at a given nucleotide and then dividing the height of the G peak by the sum of heights of the A and G peaks at a specific site. Measurement of peak heights can be done manually using a program such as Adobe Photoshop or automatically using a peak-calling program such as Bio-Edit (http://www.mbio.ncsu.edu/bioedit/bioedit.html).

3.3 Identifying Editing Sites by RT-PCR and High-Throughput Sequencing

While RT-PCR followed by Sanger Sequencing is still the gold standard for editing site examination, that approach only provides information on the extent of editing at a given site within the total pool of mRNAs in the sample. However, A-to-I editing events often occur at multiple positions within one gene. To determine the co-occurrence of multiple editing events, individual transcripts must be sequenced [16–18]. To accomplish sequencing of multiple editing events on individual transcripts, RT-PCR products for each mRNA (*see* **Note 23**) can be ligated to adapters and subjected to high-throughput sequencing (Fig. 7). This method also provides a very accurate means of determining the percentage of editing at each adenosine in a given gene.

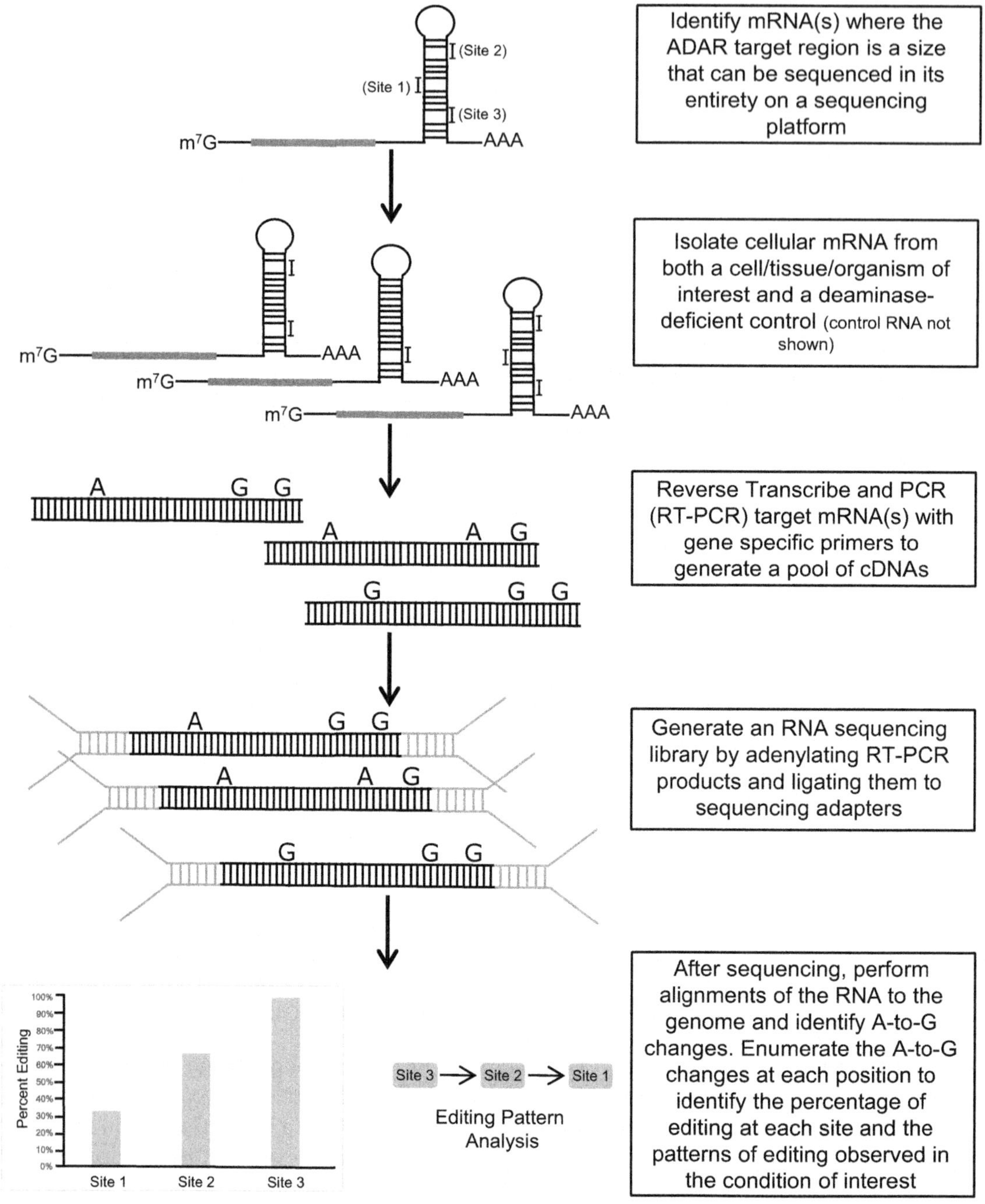

Fig. 7 Overview of RT-PCR and high-throughput sequencing method for detecting A-to-I editing events

3.3.1 Primer Design

1. The ability to sequence an entire edited region of a transcript is limited by the size of the sequencing platform. For example, at the date of this publication, the size limit for the Illumina MiSeq platform is 2 × 300 bp (http://www.illumina.com/systems/sequencing.html), which means that a mRNA region

of approximately 450–500 base-pairs could be sequenced using paired end sequencing (*see* **Note 24**). Therefore, after determining the size of the edited regions that you want to examine, confirm that size can be sequenced on an available platform.

2. Design a reverse transcription primer that lies outside the edited region and any double-stranded RNA that is predicted to occur in the mature mRNA (described in Subheading 3.2.1).

3.3.2 Reverse Transcription and RT-PCR

1. Extract the RNA (Subheading 3.1), reverse-transcribe with gene-specific primers (Subheading 3.2.3), and perform a PCR for each gene of interest (Subheading 3.2.4).
2. Subject a small portion (usually 1/10th) of the PCR reaction to agarose gel electrophoresis to confirm that the RT-PCR process amplified cDNA of the correct size from the reverse transcription reaction and that no contaminating genomic DNA is present in the RNA (*see* **Note 15**).
3. After confirming the RT-PCR, the remaining product is purified from the PCR primers using a QIAquick PCR purification kit.

3.3.3 Library Preparation

1. Determine the concentration of the PCR products accurately using a Nanodrop (>10 ng/μL) or Qubit Fluorometer (<10 ng/μL), depending upon the concentration of the sample.
2. Combine equal molar amounts of the RT-PCR products from each condition to be analyzed. Then perform an end-repair reaction by incubating the following reaction in a thermomixer at 20°C for 120 min.

1.2 μg RT-PCR products	__ μL
T4 DNA Polymerase	1 μL
1:5 diluted Klenow DNA Polymerase	1 μL
T4 Polynucleotide Kinase	1 μL
T4 DNA Ligase buffer with 10 mM ATP	5 μL
10 mM dNTPs	2 μL
Water (up to 50 μL)	__ μL

3. Stop the end-repair reaction by heat inactivation of the enzymes at 75°C for 20 min.
4. Purify the end-repaired RT-PCR using a QIAquick PCR purification kit. Elute with 35 μL of elution buffer.

5. To allow for ligation to the Illumina Tru-Seq adapters, which have a thymidine overhang, subject the end-repaired products to 3′ adenylation by incubating the following reaction at 37°C for 2 h.

End-repaired RT-PCR products	34 μL
Klenow Fragment (3′-5′ exonuclease minus)	1 μL
Klenow Buffer	5 μL
5 mM dATP	10 μL

6. Purify the 3′ adenylated products from the enzymes using a MinElute PCR Purification Kit, with the product eluted in 15 μL of elution buffer.
7. Ligate the RT-PCR products to Illumina Tru-Seq sequencing adapters by incubating the following reaction at 16 °C overnight.

Product from **step 5**	15 μL
10× DNA ligase buffer	3 μL
T4 DNA ligase	1 μL
Annealed Tru-Seq adapters (~1.5 μM)	5–10 μL
dH_2O (up to 30 μL)	__ μL

8. To remove any remaining unligated adapters from the library before sequencing, add 20 μL of water to the reaction from **step 6** and then purify using AMPure XP magnetic beads (*see* **Note 25**) per the following protocol:
   (a) Add 50 μL of AMPureXP beads and incubate at room temperature for 1 min (*see* **Note 26**).
   (b) Place the bead/library mixture in a magnetic stand for 2–3 min.
   (c) Using thin-tipped pipet tips, remove the supernatant.
   (d) Slowly pipet 70 μL of ice-cold 70% ethanol over the magnetic beads.
   (e) Turn the tube 180° in the magnetic stand and let it stand for 2–3 min.
   (f) Repeat **step c** above.
   (g) Remove the tube from the magnetic stand and add 16 μL of QIAGEN elution buffer. This should force the beads to the bottom of the tube. If not, repeat the pipetting using the elution buffer in the tube (*see* **Note 27**).
   (h) Place the tube in the magnetic stand for 3–5 min.
   (i) Remove 15 μL of supernatant (library) to a new tube.

(j) Library quality and concentration can be determined using Bioanalyzer and qRT-PCR quantification (using a commercial assay such as the KAPA library quantification kit), respectively. This can be performed before sending for sequencing to confirm a successful library preparation, but is also typically performed in the sequencing facility.

3.3.4 RNA Sequencing and Editing Site Analysis

1. To accurately determine A-to-I editing sites on each transcript and the levels of editing for each gene, each library, which represents the RT-PCR product of one (or more) genes (*see* **Note 23**) in one cell/tissue/organism, should be sequenced at a depth of ~10,000 reads per gene.
2. The resulting reads should be trimmed and scored for accuracy using a program such as Trimmomatic or cutadapt.
3. After removing low-quality reads, overlap the paired-end reads using a program such as Flash to generate a single long read spanning the double-stranded region (*see* **Note 24**).
4. Using a program such as Bowtie, align the surviving reads to the genomic sequence of the individual transcripts subjected to RT-PCR (*see* **Note 28**).
5. At this step, filters can be employed to remove single-nucleotide variants between the RT-PCR product sequences and the reference sequences that are likely the result of reverse transcription, PCR or sequencing errors and not the result of A-to-I editing. These include, but are not limited to:
   (a) Removing reads with internal deletions/amplifications.
   (b) Removing reads with non-A-to-G single nucleotide variants.
   (c) Removing reads with low-quality sequencing.
6. Identify all A-to-G changes between the aligned RT-PCR product sequences and the reference sequences. Computationally predict ADAR editing sites by comparing the sequence of the RT-PCR products from a given cell type/tissue/organism to the sequencing products of the RT-PCR products from a deaminase-deficient control. Editing site identification can be based on a minimum frequency of A-to-G changes observed in the sequencing data of the RT-PCR products [21] or by using a statistical analysis of frequency of A-to-G changes observed in the cell type/tissue/organism compared to the deaminase-deficient control [17].
7. As this sequencing data contains information on editing of individual transcripts, additional downstream analysis can include determination of the number of editing events per transcript and enrichment or depletion of specific editing site patterns for a given gene. For these analyses, the editing events

that occur on a single long read are enumerated and then that information is used to generate counts of the patterns of editing on individual transcripts. The observed patterns of A-to-I editing events can be compared to expected values for each pattern, which are calculated based on the number of reads for each library multiplied by the amount of editing at a given adenosine.

8. Identification of novel editing sites using this method should be confirmed using the RT-PCR and Sanger sequencing method described above (*see* **Note 29**).

3.4 Identifying RNA Editing Sites by Transcriptome-Wide RNA Sequencing

The techniques described thus far require knowledge of the editing site or at least the edited gene of interest before utilizing the protocol. However, many groups have sought to identify novel A-to-I editing sites that occur in a specific organism [22], in cancerous tissues [23–25], and at specific times during development [26] in order to understand the biological consequences of A-to-I RNA editing. In contrast to the previous two methods, a transcriptome-wide approach to identify novel editing sites requires an unbiased method for reverse-transcription/PCR and a robust bioinformatics approach to correctly predict A-to-I editing events in the RNA-Seq data. Here, we briefly describe a method for generating a transcriptome-wide mRNA sequencing library and some main steps/considerations in the bioinformatic analysis of A-to-I editing.

3.4.1 Enrichment of Polyadenylated mRNA Using Magnetic Oligo (dT) Beads

1. For each RNA sample to be subjected to transcriptome-wide sequencing, aliquot 75 μL of magnetic oligo (dT) beads into a 1.5 mL tube (*see* **Note 30**).
2. Place the tube on a magnetic stand and remove buffer.
3. Add 100 μL Binding buffer and turn tube 180° on a magnetic stand (*see* **Note 26**). After 1–2 min, remove the supernatant carefully with thin-tipped pipet tips.
4. Repeat **step 3** two additional times. Then, resuspend beads in 50 μL Binding buffer. Place beads on ice until needed.
5. Resuspend RNA (≤10 μg) in 50 μL water and add 50 μL Binding buffer.
6. Heat RNA at 65°C for 5 min. Then place on ice for 1 min (*see* **Note 17**).
7. Add 100 μL of RNA mixture to the magnetic beads and place the tube in a rotator to gently spin for 5 min at room temperature.
8. Place the tube in a magnetic stand for 1–2 min. Remove the supernatant.
9. Add 200 μL Wash buffer and turn the tube 180° on a magnetic stand.

10. Remove the supernatant and add 100 μL Binding buffer.
11. Turn tube 180° on a magnetic stand. Remove the supernatant and add 50 μL of 10 mM Tris–HCl, pH 7.8 that has been heated to 80°C.
12. Place the tube at 80°C for 2 min and then return to a magnetic stand for 1–2 min.
13. Pipet supernatant (one round of polyA+ selected RNA) into a new tube on ice, which contains 50 μL of chilled Binding buffer.
14. Wash eluted magnetic beads two times with 200 μL Wash buffer, making sure to remove all remaining Wash buffer after the second wash.
15. Add 50 μL of Binding buffer to beads and place on ice.
16. Repeat **steps 6–9**.
17. Repeat **step 9** two additional times, making sure to remove all remaining Wash buffer after the third wash.
18. Depending upon library preparation kit to be used, either elute the polyA$^+$ RNA from the magnetic beads using 16 μL 10 mM Tris–HCl, pH 7.8 or add fragmentation buffer (supplied with some commercial library preparation kits) directly to magnetic beads.

3.4.2 Library Preparation

Since A-to-I editing events occur in double-stranded regions, there are two critical steps during the preparation of the sequencing library that need to be included to ensure both thorough and accurate identification of A-to-I editing sites across the transcriptome (Fig. 8). The first is to include an RNA fragmentation step before reverse transcription. This will allow the highly double-stranded regions that are edited by ADARs to be more efficiently transcribed by the reverse transcriptase, thus increasing the sequencing coverage in these transcriptomic regions. The second critical step is to utilize a library preparation protocol that allows the sequencing reads to be specifically assigned to a gene/genic region transcribed from the sense or antisense strand of the genome. Since the double-stranded regions that are edited by ADARs are often comprised of two inverted repetitive elements that base-pair, it is necessary to maintain the identity of the strand during sequencing in order to assign the reads and editing events specifically to one or the other repeat. Strand identity is often maintained in library preparation approaches by using dUTP in place of dTTP during second strand synthesis of the cDNA and then using Uracil DNA glycosylase immediately before PCR amplification of the library. There are many different commercial library preparation kits that fit these two criteria and can be utilized by following manufacturer's instructions.

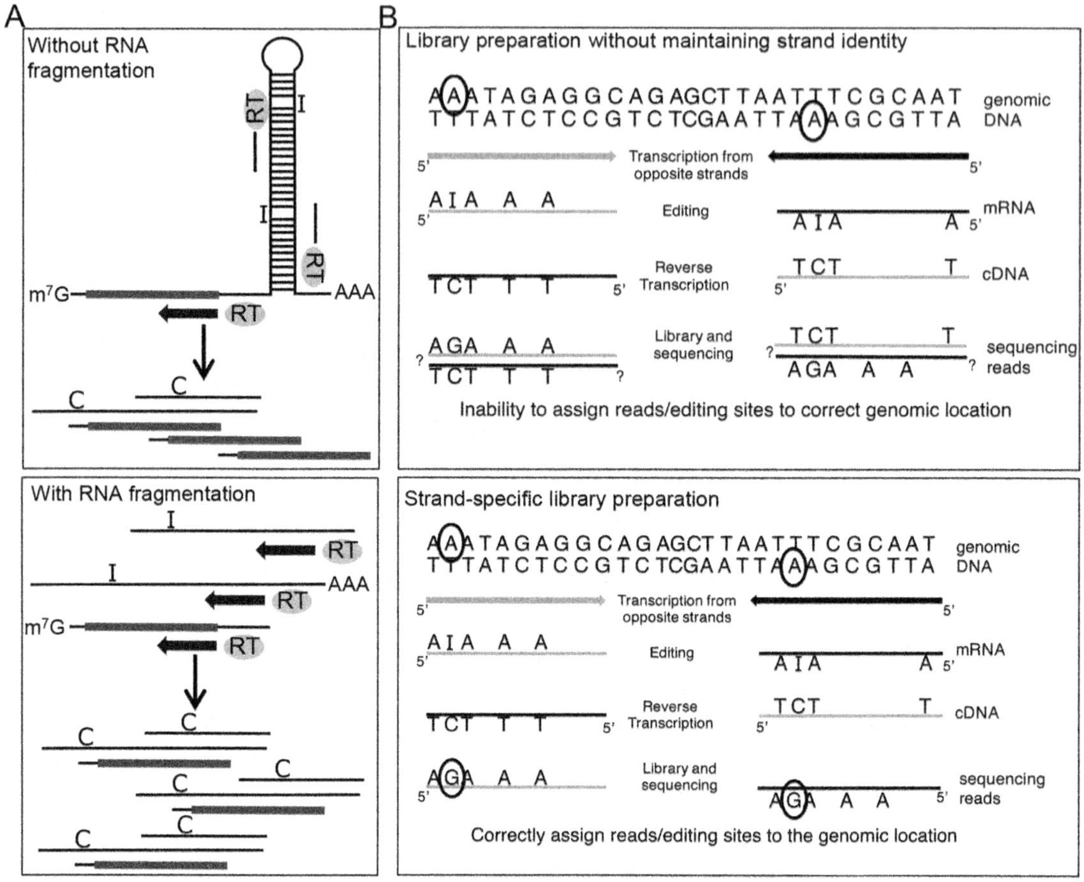

Fig. 8 Critical aspects to consider for transcriptome-wide analysis of A-to-I editing. (**a**) Top panel: ADARs target double-stranded regions of RNA, which are inefficiently transcribed into cDNA by reverse transcriptase (RT). Bottom panel: RNA fragmentation prior to reverse transcription removes the double-stranded barriers from reverse transcriptase (RT). (**b**) mRNAs can be synthesized from either strand of genomic DNA and have similar sequences. Top panel: If the strand identity is not maintained during the library preparation, the resulting sequencing reads will be identical. Identical reads would align to both locations in the genome, thus, the correct assignment of the editing site would not be possible. Bottom panel: If the library preparation includes a step that maintains strand identity, the resulting sequencing reads will differ in directionality. This difference will allow the reads to be aligned to either the sense or antisense strand of the genome and thus, correct identification of the editing site

3.4.3 RNA Sequencing and Editing Site Analysis

1. To accurately identify A-to-I editing sites, each library—which represents the entire transcriptome of one cell type/tissue/organism—should be sequenced at a depth of at least 5–10 reads (*see* **Note 30**) for all parts of the transcript. As this varies depending upon the number of transcripts expressed in a given cell type/tissue/organism, the overall read depth recommendation can vary from 10 to 100 million reads per library.

2. Trim the resulting reads using a program such as Trimmomatic or cutadapt.
3. Using a program such as Bowtie or STAR, align the surviving reads to a reference genomic sequence.
4. At this step, filters can be employed to remove single-nucleotide variants between the amplified cDNA sequences and the reference sequences that are likely the result of reverse transcription, PCR or sequencing errors and not the result of A-to-I editing. These include, but are not limited to:
   (a) Removing reads with internal deletions/amplifications.
   (b) Removing reads with non-A-to-G single-nucleotide variants.
   (c) Removing reads with low-quality sequencing.
   (d) Removing duplicated reads.
   (e) Removing reads that do not overlap a splice junction by at least 10 nucleotides (nt) on either side.
   (f) Removing reads with single-nucleotide variants within 5 nt from end.
5. Identify all A-to-G variants between the aligned RNA sequencing data and the reference sequence using programs such as Samtools mpileup and bcftools. These candidate editing sites can then be further classified by assigning a confidence score using a method such as the Bayesian model described in [27]. Typically, sites with a confidence score above 99% are considered true ADAR editing sites (*see* **Note 31**).
6. Identification of several novel editing sites using this method should be confirmed using the RT-PCR and Sanger sequencing method described above to determine the false discovery rate of the pipeline (*see* **Note 29**).

4 Notes

1. All the steps including weighing of chemicals need to be performed in an RNase-free environment. Bake all glassware, metal stir bars, and metal spatulas at 200°C for at least 24 h to reduce chances of contamination.
2. Water that is utilized for RNA is taken directly from a still or a Millipore water purification system, placed into glassware that was baked at 200°C and autoclaved. While some protocols suggest treating water with diethyl pyrocarbonate (DEPC) for use with RNA; DEPC can react with adenosines in RNA and thus skew RNA editing analysis.

3. Klenow Fragment buffer is equivalent to NEB Restriction enzyme Buffer #2.
4. Cells and small organisms, such as *C. elegans*, can be lysed by the freeze/thaw/vortex cycles described in this step. For tissue samples, the tissue should be placed in a tube on dry ice, TRIzol should be added and then the tissue homogenized with a disposable mortar. After this homogenization, the tissue/TRIzol mixture is also subjected to three freeze/thaw/vortex cycles.
5. To accurately determine editing levels, it is critical to isolate RNA free from DNA contamination. This step is most likely to result in contamination; therefore, when removing the supernatant containing RNA, very cautiously avoid the interface between the clear and white layers. The reduction in RNA yield from avoiding this layer is minimal, especially considering the lack of utility of RNA contaminated with DNA.
6. For RNA concentrations less than 50 μg, TURBO DNase can be reduced to 2.5 μL.
7. Editing targets in coding regions tend to be within double-stranded regions of pre-mRNA, but these structures are removed upon splicing. As editing analysis will be performed on mature mRNA, folding of the mature mRNA around a predicted editing event in a coding region may not yield a double-stranded structure.
8. Primers are best when between 25 and 30 bp long, start and end in GC-rich regions and do not contain stretches of three or more bases of self-annealing or complementary regions. Design primers such that they do not fall in double-stranded regions.
9. Primers within the double-stranded region of the cDNA do not anneal well to their targets as primer binding to the cDNA is competed by the cDNA intramolecularly base-pairing. The region amplified should ideally be 1000 bp or less.
10. The second PCR amplification needs to be done with a different primer pair than the first PCR to avoid further amplifying any nonspecific products that may have been amplified in the first PCR.
11. The sequencing primer should ideally be not more than 500 bp away from last editing site to obtain best results. In addition, the sequence obtained within the first 50 bp of the primer is usually too noisy to identify A-to-I editing sites. One of the 2nd PCR primers can be used as the sequencing primer as well. Last, if there are homopolymer stretches of five nucleotides or greater, continuous sequencing in those regions will be difficult and may require two sequencing reactions with primers surrounding those areas.
12. A temperature gradient from 45°C to 60°C is sufficient to determine optimal annealing temperature.

13. If an editing event is in the coding region and the primers surround an intronic sequence, genomic DNA cannot be used to optimize annealing temperature of the PCR.
14. If concerned about the specificity of PCR amplification, the products from the annealing optimization PCR can be subjected to Sanger sequencing.
15. A minus reverse transcriptase (RT) sample for each RNA sample that is reverse transcribed must also be run to ensure that there is no DNA contamination in the RNA. *See* **Note 6** above.
16. If the target double-stranded region is small, Superscript III can be used in lieu of Thermoscript reverse transcriptase.
17. Do not allow samples to sit on ice for longer than 1 min.
18. Do not allow samples to remain at 85°C longer than 10 min.
19. If the PCR product is 1.5 kb or larger, add another 1 μL of dNTPs. Also, genomic DNA only serves as positive control if primers do not span an exon-intron junction.
20. The amount of first PCR product used as a template in the second PCR reaction can vary from 0.5 μL to 3 μL. However, using more than 3 uL does not typically yield better results. If you still have no product, re-optimize the first PCR reaction.
21. Quantification of editing levels by Sanger sequencing requires high-quality chromatograms that have sharp peaks and very little noise. To obtain these types of chromatograms, it is important to precisely quantify the amount of RT-PCR product that will be placed into the Sanger sequencing reaction. In addition, the size of the RT-PCR product influences how much template to place into the reaction. The recommended amounts of template in sequencing reaction are as follows: 10–20 ng for a 500 bp PCR product, 20–40 ng for a 1 kb PCR product, 40–80 ng for a 2 kb PCR product.
22. The G peak must be taller than the background noise of neighboring peaks to be considered an editing event.
23. The transcripts from multiple genes can be simultaneously analyzed using this protocol. However, all the RT-PCR products need to be of a similar length. For example, if sequencing a 300 bp region of a transcript, all RT-PCR products should be within 25–50 bp of that target length.
24. As sequencing quality decreases over the length of the read and the goal of this method is to determine all the editing sites on an individual transcript, this method is most accurate when paired-end sequencing is utilized.
25. Before starting the clean-up step, the AMPure XP beads need to be at room temperature.
26. To minimize loss during the bead purification, the lid of the tube is kept open.

27. Do not pipet library/AMPure XP bead mixture up and down repeatedly.
28. As A-to-I editing events can occur in clusters [11], each sequencing read may have multiple editing sites. To allow these reads to align to the genomic reference sequences, the alignment needs to be performed allowing for multiple mismatches. While in transcriptome-wide sequencing allowing multiple mismatches can result in improper mapping of reads to the genome [19], this targeted sequencing approach only aligns the reads to the genomic reference sequences subjected to RT-PCR and thus mis-mapping is unlikely to occur. Thus, to obtain the most information about editing on a given transcript, it is recommended to allow for at least ten mismatches per read in the alignment programs.
29. A given A-to-I editing event needs to be edited in approximately 5–10% of transcripts to be detected using the RT-PCR and Sanger sequencing method.
30. Single-read (also referred to as single end) sequencing is sufficient for accurately identifying editing sites in transcriptome-wide RNA-Seq experiments.
31. The parameters for computationally predicting ADAR editing sites are facilitated by running the same bioinformatics pipeline on RNA sequencing data from a deaminase-deficient control and identifying no or very few A-to-G changes. Any A-to-G changes in this dataset can be subtracted from the computationally predicted A-to-I editing sites for the cell type/tissue/organism of interest, thus increasing the accuracy of the editing site identification.

Acknowledgment

Work in the Hundley lab is supported by Research Scholar Award RSG-15-051-01-RMC from the American Cancer Society. We thank Aidan Manning and Dr. Sarah Deffit for helpful comments on the manuscript.

References

1. Gott JM, Emeson RB (2000) Functions and mechanisms of RNA editing. Annu Rev Genet 34:499–531. doi:10.1146/annurev.genet.34.1.499
2. Licht K, Jantsch MF (2016) Rapid and dynamic transcriptome regulation by RNA editing and RNA modifications. J Cell Biol 213(1):15–22. doi:10.1083/jcb.201511041
3. Mannion N, Arieti F, Gallo A, Keegan LP, O'Connell MA (2015) New insights into the biological role of mammalian ADARs; the RNA editing proteins. Biomol Ther 5(4):2338–2362. doi:10.3390/biom5042338
4. Savva YA, Rieder LE, Reenan RA (2012) The ADAR protein family. Genome Biol 13(12):252. doi:10.1186/gb-2012-13-12-252

5. Samuel CE (2012) ADARs: viruses and innate immunity. Curr Top Microbiol Immunol 353:163–195. doi:10.1007/82_2011_148
6. Gommans WM (2012) A-to-I editing of microRNAs: regulating the regulators? Semin Cell Dev Biol 23(3):251–257. doi:10.1016/j.semcdb.2011.09.018
7. Daniel C, Lagergren J, Ohman M (2015) RNA editing of non-coding RNA and its role in gene regulation. Biochimie 117:22–27. doi:10.1016/j.biochi.2015.05.020
8. Nishikura K (2016) A-to-I editing of coding and non-coding RNAs by ADARs. Nat Rev Mol Cell Biol 17(2):83–96. doi:10.1038/nrm.2015.4
9. Li JB, Levanon EY, Yoon JK, Aach J, Xie B, Leproust E, Zhang K, Gao Y, Church GM (2009) Genome-wide identification of human RNA editing sites by parallel DNA capturing and sequencing. Science 324(5931):1210–1213. doi:10.1126/science.1170995
10. Peng Z, Cheng Y, Tan BC, Kang L, Tian Z, Zhu Y, Zhang W, Liang Y, Hu X, Tan X, Guo J, Dong Z, Liang Y, Bao L, Wang J (2012) Comprehensive analysis of RNA-Seq data reveals extensive RNA editing in a human transcriptome. Nat Biotechnol 30(3):253–260. doi:10.1038/nbt.2122
11. Porath HT, Carmi S, Levanon EY (2014) A genome-wide map of hyper-edited RNA reveals numerous new sites. Nat Commun 5:4726. doi:10.1038/ncomms5726
12. Ramaswami G, Lin W, Piskol R, Tan MH, Davis C, Li JB (2012) Accurate identification of human Alu and non-Alu RNA editing sites. Nat Methods 9(6):579–581. doi:10.1038/nmeth.1982
13. Ramaswami G, Zhang R, Piskol R, Keegan LP, Deng P, O'Connell MA, Li JB (2013) Identifying RNA editing sites using RNA sequencing data alone. Nat Methods 10(2):128–132. doi:10.1038/nmeth.2330
14. Slotkin W, Nishikura K (2013) Adenosine-to-inosine RNA editing and human disease. Genome Med 5(11):105. doi:10.1186/gm508
15. Tariq A, Jantsch MF (2012) Transcript diversification in the nervous system: a to I RNA editing in CNS function and disease development. Front Neurosci 6:99. doi:10.3389/fnins.2012.00099
16. Enstero M, Daniel C, Wahlstedt H, Major F, Ohman M (2009) Recognition and coupling of A-to-I edited sites are determined by the tertiary structure of the RNA. Nucleic Acids Res 37(20):6916–6926. doi:10.1093/nar/gkp731
17. Wheeler EC, Washburn MC, Major F, Rusch DB, Hundley HA (2015) Noncoding regions of *C. elegans* mRNA undergo selective adenosine to inosine deamination and contain a small number of editing sites per transcript. RNA Biol 12(2):162–174. doi:10.1080/15476286.2015.1017220
18. Veno MT, Bramsen JB, Bendixen C, Panitz F, Holm IE, Ohman M, Kjems J (2012) Spatio-temporal regulation of ADAR editing during development in porcine neural tissues. RNA Biol 9(8):1054–1065. doi:10.4161/rna.21082
19. Bass B, Hundley H, Li JB, Peng Z, Pickrell J, Xiao XG, Yang L (2012) The difficult calls in RNA editing. Interviewed by H Craig Mak. Nat Biotechnol 30(12):1207–1209. doi:10.1038/nbt.2452
20. Zuker M (2003) Mfold web server for nucleic acid folding and hybridization prediction. Nucleic Acids Res 31(13):3406–3415
21. Bazak L, Haviv A, Barak M, Jacob-Hirsch J, Deng P, Zhang R, Isaacs FJ, Rechavi G, Li JB, Eisenberg E, Levanon EY (2014) A-to-I RNA editing occurs at over a hundred million genomic sites, located in a majority of human genes. Genome Res 24(3):365–376. doi:10.1101/gr.164749.113
22. Alon S, Garrett SC, Levanon EY, Olson S, Graveley BR, Rosenthal JJ, Eisenberg E (2015) The majority of transcripts in the squid nervous system are extensively recoded by A-to-I RNA editing. eLife 4. doi:10.7554/eLife.05198
23. Fumagalli D, Gacquer D, Rothe F, Lefort A, Libert F, Brown D, Kheddoumi N, Shlien A, Konopka T, Salgado R, Larsimont D, Polyak K, Willard-Gallo K, Desmedt C, Piccart M, Abramowicz M, Campbell PJ, Sotiriou C, Detours V (2015) Principles governing A-to-I RNA editing in the breast cancer transcriptome. Cell Rep 13(2):277–289. doi:10.1016/j.celrep.2015.09.032
24. Paz-Yaacov N, Bazak L, Buchumenski I, Porath HT, Danan-Gotthold M, Knisbacher BA, Eisenberg E, Levanon EY (2015) Elevated RNA editing activity is a major contributor to transcriptomic diversity in tumors. Cell Rep 13(2):267–276. doi:10.1016/j.celrep.2015.08.080
25. Han L, Diao L, Yu S, Xu X, Li J, Zhang R, Yang Y, Werner HM, Eterovic AK, Yuan Y, Li J, Nair N, Minelli R, Tsang YH, Cheung LW, Jeong KJ, Roszik J, Ju Z, Woodman SE, Lu Y, Scott KL, Li JB, Mills GB, Liang H (2015) The genomic landscape and clinical relevance of A-to-I RNA editing in human cancers. Cancer Cell 28(4):515–528. doi:10.1016/j.ccell.2015.08.013
26. Qiu S, Li W, Xiong H, Liu D, Bai Y, Wu K, Zhang X, Yang H, Ma K, Hou Y, Li B (2016) Single-cell RNA sequencing reveals dynamic changes in A-to-I RNA editome during early human embryogenesis. BMC Genomics 17(1):766. doi:10.1186/s12864-016-3115-2
27. Washburn MC, Kakaradov B, Sundararaman B, Wheeler E, Hoon S, Yeo GW, Hundley HA (2014) The dsRBP and inactive editor ADR-1 utilizes dsRNA binding to regulate A-to-I RNA editing across the *C. elegans* transcriptome. Cell Rep 6(4):599–607. doi:10.1016/j.celrep.2014.01.011

Chapter 10

Using RNA-Seq to Discover Genetic Polymorphisms That Produce Hidden Splice Variants

Shayna Stein, Emad Bahrami-Samani, and Yi Xing

Abstract

RNA-seq is a powerful and popular technology for studying posttranscriptional regulation of gene expression, such as alternative splicing. The first step in analyzing RNA-seq data is to map the sequenced reads back to the genome. However, commonly used RNA-seq aligners use the consensus splice site dinucleotide motifs to map reads across splice junctions. This can be deceiving due to genomic variants that create novel splice site dinucleotides, leaving the personal splice junction reads un-mapped to the reference genome. We developed and evaluated a method called RNA Personal Genome Alignment Analyzer (rPGA) to identify "hidden" splicing variations in personal transcriptomes, by mapping personal RNA-seq data to personal genomes. Our work demonstrates that the personal genome approach to RNA-seq read alignment enables the discovery of a large but previously unknown catalog of splicing variations in human populations.

Key words Alternative splicing, RNA-seq, Single-nucleotide polymorphisms (SNPs)

1 Introduction

With emerging personalized genomics we are entering the era of genome-guided medicine, moving us toward the modern age of genetics with more effective medical interventions. Owing to new technologies and computational methods, we now know a great deal about mechanisms, with which an alteration in the genome can cause a disease by affecting gene expression regulatory processes. Despite a complex and multifaceted nature, posttranscriptional regulation of gene expression is no exception in such mechanisms, and genetic variation causing aberrations in posttranscriptional gene regulation can have significant disease causing consequences. Perhaps the most important of all is alternative splicing (AS), the process in which exons are differentially included in the mature mRNA products during pre-mRNAsplicing [1]. It is a predominant mechanism for producing multiple distinct mRNA and protein isoforms from a single gene [2].

Yongsheng Shi (ed.), *mRNA Processing: Methods and Protocols*, Methods in Molecular Biology, vol. 1648,
DOI 10.1007/978-1-4939-7204-3_10, © Springer Science+Business Media LLC 2017

Recent studies have discovered numerous single-nucleotide polymorphisms (SNPs) associated with nearby splice variants. These SNPs, referred to as splicing quantitative trait loci (sQTLs) [3–5], may affect resulting transcript isoforms produced from a gene, which in turn contribute to natural transcriptomic and phenotypic diversity in the human population. Identifying sQTLs may elucidate underlying mechanisms that influence disease risk and treatment outcomes [6–8]. AS is partially regulated by highly conserved splicing signals within the 5′ and 3′ splice sites. Approximately 99% of mammalian splice sites follow the "GT-AG" rule, which states that the first two and last two nucleotides in the intron are GT and AG, respectively [9]. SNPs that create or corrupt the dinucleotide motif likely alter splicing patterns and the resulting transcript isoforms [10]. In fact, such mutations represent a large class of disease mutations [11]. Thus, discovering splice variants will likely to reveal molecular mechanisms regulating disease-related AS.

A powerful technology to study AS is RNAsequencing (RNA-seq) [2]. The first step of the computational analysis of the resulting RNA-seq data involves aligning sequenced fragments of mRNA (i.e., reads) to a reference genome. Reads aligning to exons and splice junctions are then counted to quantify AS events [12–15]. Traditional RNA-seq alignment relies on the human reference genome; however, this has important limitations when analyzing individual variations oftranscriptomes. One such issue is that commonly used alignment algorithms, including STAR [16] and Tophat [17], rely on the consensus splice site dinucleotide motif to identify splice junctions. Therefore, if a genetic variant creates a novel splice site motif the resulting splice junction reads are likely to be unmappable to the reference genome, causing the splice junction to be missed, as shown in Fig. 1.

Motivated by this problem, we have developed a personal genome approach to RNA-seq alignment to enable the discovery of previously unknown splicing variations. Our computational pipeline, rPGA [18], personalizes the reference genome according to an individual's SNPs, and then maps the individual'stranscriptome to the corresponding personal genome, and discovers novel splice variants specific to the individual, as shown in Fig. 2. Details of the rPGA pipeline and its application are described previously [18]. We additionally have a second pipeline designed to provide allele-specific RNA-seq assignments (Fig. 3) that can be used to analyze other allelic effects of genetic variants on RNA regulation, and details of the allele assignments are provided in Subheading3.3.3. rPGA requires four input files: the RNA-seq data [raw sequence files (FASTQ)], the genotype file (VCF), a reference genome sequence (FASTA), and a gene/transcript annotation file (GTF). This pipeline can be used on any eukaryotic organism with a complete reference genome and gene/transcript annotations.

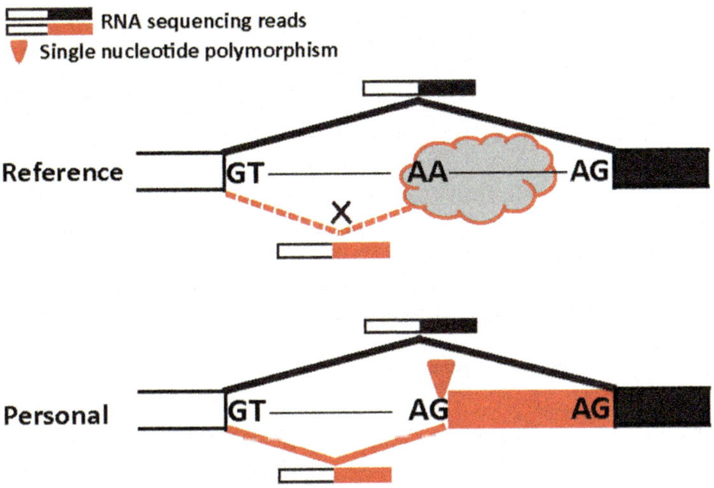

Fig. 1 Hidden splice variants in RNA-seq read alignment. RNA-seq splice junction reads originating from SNPs creating personal splice site dinucleotide motifs (shown in *red*) do not align the reference genome due to noncanonical splice site motifs in the reference genome. The RNA-seq splice junction reads do, however, align to the personal genome. Figure adapted from [18]

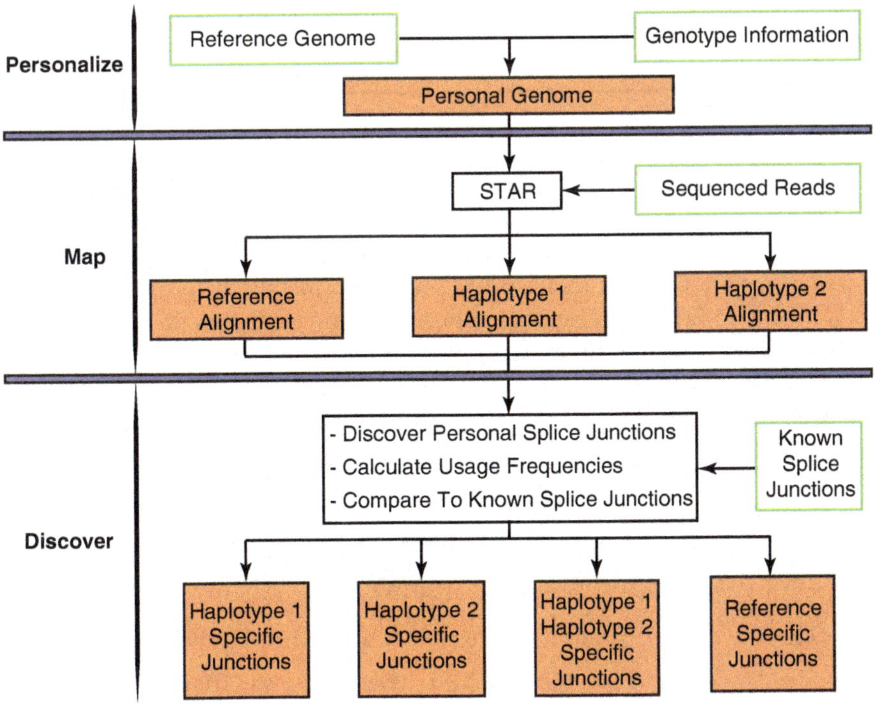

Fig. 2 Flowchart of the rPGA pipeline to identify hidden splice variants in personaltranscriptomes. Figure adapted from [18]

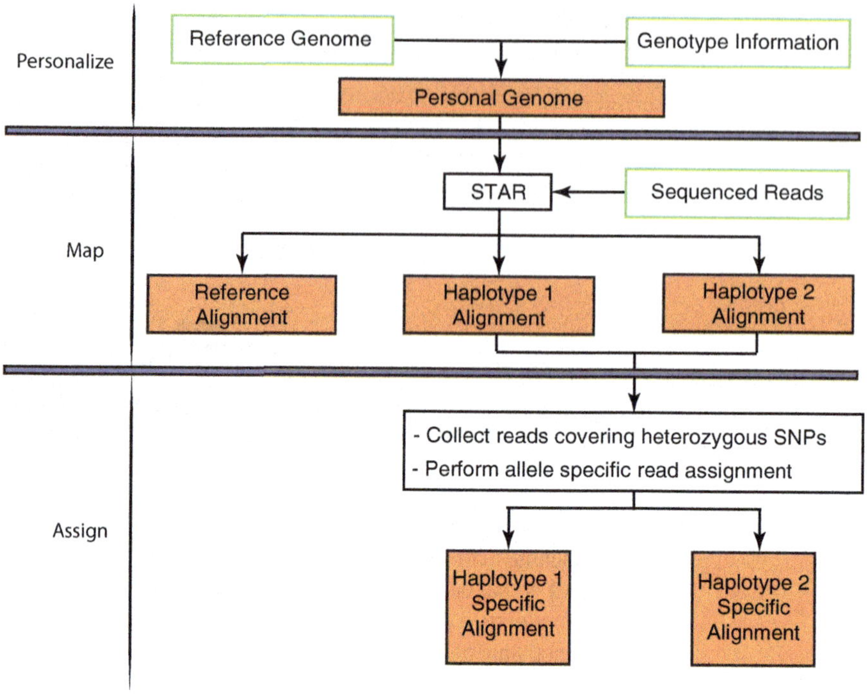

Fig. 3 Flowchart of the rPGA pipeline to identify allele-specific RNA-seq reads

2 Materials

rPGA is intended to be used in a Unix-based environment. It has been tested on Mac OS and Linux. A working installation of Python is required. rPGA makes use of the STAR software package for alignment. STAR must be installed. By default, rPGA expects them to be somewhere in your path, but if they are not you can specify their locations when running the configure script.

2.1 Prerequisites

1. Download and install Python 2.7x (https://www.python.org/downloads/release/python-2711).
2. Download and install SAMtools (https://samtools.github.io/).
3. Download and install STAR (version 2.4.2a, https://github.com/alexdobin/STAR) (*see* **Note 1**).

2.2 Install rPGA

1. Install prerequisites described in Subheading 2.1.
2. Add the Python directory to the $PATH environment variable.
3. Add the STAR directories to the $PATH environment variable.
4. Add the SAMtools directory to the $PATH environment variable.

5. Download the most recent version of rPGA (https://github.com/Xinglab/rPGA).
6. Obtain the human reference genome (hg19) (http://hgdownload.cse.ucsc.edu/goldenPath/hg19/chromosomes/). Concatenate chromosome FASTA files into a single file (cat *.fa > hg19.fa).
7. Obtain the RNA-seq reads and corresponding VCF file for your particular sample of interest.
8. Unpack rPGA downloaded from**step 5** and cd into the newly created directory:

```
tar -xf rPGA-0.0.1.tar.gz
cd rPGA
```

9. Configure the rPGA Installation

```
./configure
```

   rPGA is written partially in Python and will use whichever python interpreter is invoked whenever you type "Python" on the command line. If you wish to specify a different Python version, specify its location when running configure. Furthermore, if any of the required external software packages listed above is not in your path, you can specify their locations when running configure. For example, to specify the path to Python and STAR you would type the following:

```
./configure --with-python=/path/to/python
-with-STAR=/path/to/STAR
```

   You need only to include the ones that cannot be found in your path. The configure script will tell you if some requirements cannot be found.
10. Install rPGA. User executables and scripts will be placed into /path/to/rPGA/bin and /path/to/rPGA/scripts. These cannot be moved, as the pipeline expects them to be there. We recommend you to add these directories to your path, and the following instructions assume they are; if not, replace the names of the scripts and executables with their full path.

3 Methods

rPGA both identifies hidden splice variants and performs allele specific RNA-seq alignment. Splice variants and allele specific reads are identified from the RNA-seq data, as shown in Figs. 2 and3 respectively.

3.1 Input Data Format

- Read sequence file in the FASTQ format for the sequence of interest.
- Genotype file in the VCF format for the sample of interest.
- Reference file in the FASTA format for the species of interest.
- Annotation of genes and transcripts in the GTF format (*see* **Note 2**).

Once all input data are ready, run rPGA using one of the two following pipelines.

3.2 Using rPGA for Identifying Hidden Splice Variants

This is the pipeline to discover splice junctions that are hidden when aligning an individual's transcript reads to the hg19 reference genome. There are three steps in this pipeline: personalizing the genome, aligning reads, and discovering hidden splice junctions.

3.2.1 Personalize Reference Genome

- Inputs:
  1. Reference genome (FASTA).
  2. Directory containing VCF files, one per chromosome, OR a single VCF file containing all chromosomes.
- Make the personalized genome:

```
$ rPGA personalize -r reference.fa -v genotype directory -o
output_directory
```

- Outputs:
  1. Output directory/hap1.fa.
  2. Output directory/hap2.fa.
- For a complete list of options, please *see* Table 1.

3.2.2 RNA-Seq Alignment

The next step is to align the RNA-seq data to the personalized genomes using the STAR alignment tool.

- Inputs:
  1. Reference genome (FASTA).
  2. Read sequences (FASTQ), single or paired end.
  3. Personal genomes (FASTA) from personalize step.

  Please note that we choose STAR for faster alignment, but that obviously comes at a cost, and that is the required memory. Make sure you have enough memory on your computer that is running the STAR (32 GB, depending on the options of the mapper).
- Align reads to the personal genomes:

```
$ rPGAmapping -r reference.fa -s
reads 1.fastq,[reads 2.fastq] -o outdir
```

Table 1
Options and required parameters for rPGA personalize in the splice variants discovery pipeline

Parameter	Description
-o[a]	Output directory
-r[a]	Reference genome
-v[a]	VCF directory or file
–gz	Flag denoting VCF files are gzipped
–rnaedit	Flag to N-mask RNAediting sites
-e[b]	File containing RNA editing sites

[a] Required parameters
[b] Note: if –rnaedit flag is used, a file containing RNA editing events must be provided using -e. In this case, rPGA will disregard heterozygous SNPs that overlap RNA editing sites when assigning mapped reads to haplotypes

- Outputs:
  1. Output directory/HAP1/STARalign/Aligned.out.sorted.bam.
  2. Output directory/HAP1/STARalign/Aligned.out.sorted.bam.bai.
  3. Output directory/HAP2/STARalign/Aligned.out.sorted.bam.
  4. Output directory/HAP2/STARalign/Aligned.out.sorted.bam.bai.
  5. Output directory/REF/STARalign/Aligned.out.sorted.bam.
  6. Output directory/REF/STARalign/Aligned.out.sorted.bam.bai.
- For a complete list of options, please *see* Table 2.

3.2.3 Discover Hidden Splice Junctions

The final step is to discover hidden splice junctions

- Inputs:
  1. Annotation file (GTF).
  2. VCF file or directory containing VCF files, one chromosome per file.
- Usage:

```
$ rPGA discover -g annotation.gtf -v genotype directory -o
output
```

Table 2
Options and required parameters for rPGAmapping in the splice variants discovery pipeline

Parameter	Description
-o[a]	Output directory
-r[a]	Reference genome
-s[a]	Read sequences, either single- or paired-end
-T	Number of threads STAR uses, default is 8
-M	Max number of multiple alignments
-N	Max number of read mismatches, default is 3
–gz	Flag denoting sequence reads are gzipped

[a] Required parameters

- Outputs (per chromosome) Haplotype-specific bed files:
  1. hap1.chrom.specific.bed.
  2. hap2.chrom.specific.bed.
  3. hap1hap2.chrom.specific.bed.
  4. ref.chrom.specific.bed.

  Columns of each bed file are:

  1. Chrom.
  2. Junction start.
  3. Junction end.
  4. Name.
  5. Strand.
  6. Splice site usage frequency.

  The name of each splice junction is in the format J R/NC/N3/N5/N35 SNPid.

– R = junction is in the provided reference annotation.
– NC = novel combination of reference 5' and 3′ splice sites.
– N3 = novel 3′SS and reference 5′SS.
– N5 = novel 5′SS and reference 3′SS.
– N35 = both 5′SS and 3′SS are novel.
– SNPid = comma deliminated list of the splice site SNP ids, which match the SNP ids in the given VCF file.
– For a complete list of options, please*see* Table 3.

Table 3
Options and required parameters for rPGA discover

Parameter	Description
-o[a]	Output directory
-g[a]	Gene and transcript annotation file (GTF)
-v[a]	Genotype files or directory (VCF)
–gz	Flag denoting sequence reads are gzipped
–rnaedit[b]	Flag to consider RNA editing sites
-e[b]	File containing RNAediting sites
-b1[c]	Haplotype 1 alignment file to personal genome (BAM)
-b2[c]	Haplotype 2 alignment file to personal genome (BAM)
-br[c]	Reference alignment file to reference genome (BAM)

[a] Required parameters
[b] Note: if –rnaedit flag is used, a file containing RNAediting events must be provided using -e. In this case, rPGA will disregard heterozygous SNPs that overlap RNA editing sites when assigning mapped reads to haplotypes
[c] Use these if you would like to supply your own personal genome mapping alignment files. If rPGA runmapping (previous step) is used, there is no need to provide rPGA with the alignment files, and rPGA will use the alignments in HAP1/STARalign, HAP2/STARalign, REF/STARalign

3.3 Using rPGA for Allele-Specific RNA-Seq Read Alignment

To generate the allele-specific bam files, we first collect RNA-seq reads that cover each heterozygous SNP provided in the VCF files. For each read we check the position in the read corresponding to the heterozygous SNP to assign the read to either haplotype 1 or haplotype 2. The read is assigned to haplotype 1 if (1) the base corresponding to the SNP position in the read matches the hap1 SNP allele, and (2) the edit distance to the haplotype 1 personal genome is less than the edit distance to the haplotype 2 personal genome. Likewise, the read is assigned to haplotype 2 if (1) the base corresponding to the SNP position in the read matches the hap2 SNP allele, and (2) the edit distance to the haplotype 2 personal genome is less than the edit distance to the haplotype 1 personal genome. If a read covers multiple heterozygous SNPs, a majority vote is used. For example, if a read covers 3 heterozygous SNPs and 2 match the hap1 allele and 1 matches the hap2 allele AND the edit distance to hap1 is less than edit distance to hap2, the read is assigned to hap1. If a read cannot be assigned to either hap1 or hap2 according to the above rules, it is considered "conflicting" and is not assigned to either haplotype.

There are three steps in this pipeline: personalizing the genome, aligning RNA-seq reads to personal genomes, and assigning allele-specific reads.

3.3.1 Personalize Reference Genome

- Inputs:
  1. Reference genome (FASTA).
  2. Directory containing VCF files, one per chromosome, OR a single VCF file containing all chromosomes.
- Make the personalized genome.

```
$ rPGA personalize -r reference.fa -v genotype directory -o
output_directory
```

- Output (no –nmask flag):
  1. Output_directory/hap1.fa (hap1 personal genome).
  2. Output_directory/hap2.fa (hap2 personal genome).
- Output (with –nmask flag):
  1. Output_directory/nmask.fa (n-masked personal genome)
- For a complete list of options, please *see* Table 4.

Table 4
Options and required parameters for rPGA personalize in the allele-specific assignment pipeline

Parameter	Description
-o[a]	Output directory
-r[a]	Reference genome (FASTA)
-v[a]	Genotype files or directory (VCF)
–gz	Flag denoting VCF files are gzipped
–rnaedit[b]	Flag to consider RNAediting sites
-e[b]	File containing RNA editing sites
–nmask	Flag to n-mask variant sites (if using unphased data)

[a] Required parameters

[b] Note if –rnaedit flag is used, RNA editing file must be provided using -e. rPGA will change each RNA editing site to an "N" in the personal genomes. The number and locations of RNA editing sites that overlap SNPs will be reported in report.personalize.txt.

[c] Use –nmask flag to nmask SNP positions in the reference genome. This should be used if you have unphased genotype data. This produces one N-masked personal genome instead of two personal genomes. Note, –nmask option must be used for all three rPGA run steps (personalize, mapping, and alleles)

3.3.2 RNA-Seq Alignment

The next step is to alignsequencing data to the personalized genomes using the STAR alignment tool.

- Inputs:
  1. Read sequences (FASTQ).
  2. Reference genome (FASTA).

  Please note that we choose STAR for faster alignment, but that obviously comes at a cost, and that is the required memory. Make sure you have enough memory on your computer that is running the STAR (32 GB, depending on the options of the mapper).

- Align reads to the personal genomes:

```
$ rPGAmapping alleles -r reference.fa -s
reads 1.fastq,[reads 2.fastq] -o outdir
```

- Output (no –nmask flag):
  1. Output_directory/HAP1/STARalign/Aligned.out.sorted.bam.
  2. Output_directory/HAP1/STARalign/Aligned.out.sorted.bam.bai.
  3. Output_directory/HAP2/STARalign/Aligned.out.sorted.bam.
  4. Output_directory/HAP2/STARalign/Aligned.out.sorted.bam.bai.
- Output (using –nmask flag):
  5. Output_directory/MASK/STARalign/Aligned.out.sorted.bam.
  6. Output_directory/MASK/STARalign/Aligned.out.sorted.bam.bai.
- For a complete list of options, please*see* Table 5.

3.3.3 Allele-Specific Alignment

Generate allele specific bam files using the assign function.

- Input:
  1. VCF File or Directory Containing VCF Files.
- Assign allele-specific reads:

```
$ rPGA assign -v genotype directory -o outdir
```

- Output (−nmask flag not used):
  1. hap1.sorted.bam.
  2. hap1.sorted.bam.bai.
  3. hap2.sorted.bam.
  4. hap2.sorted.bam.bai.

Table 5
Options and required parameters for rPGAmapping in the allele specific assignment pipeline

Parameter	Description
-o[a]	Output directory
-r[a]	Reference genome (FASTA)
-s[a]	Read sequences, either single or paired-end (FASTQ)
-T	Number of threads STAR uses, default is 8
-M	Max number of multiple alignments, default is 20
-N	Max number of read pair mismatches, default is 6
–gz	Flag denoting read sequence files are gzipped
–nmask[b]	Flag to n-mask variant sites (if using unphased data)

[a] Required parameters

[b]Aligns RNA-seq reads to nmask.fa, producing just one alignment output. Note, –nmask option must be used for all three rPGA run steps (personalize, mapping, and alleles)

rPGA adds up to 4 SAM tags per read:

1. HT: denotes which haplotype the read originates from (1 or 2).
2. SP: semicolon deliminated list of heterozygous SNP positions the read covers.
3. GT: denotes whether the read contains the reference or alternate allele(s) for the SNP(s) in SP.
4. EP: semicolon deliminated list of RNAediting positions the read covers (if using -rnaedit).

- Output (using –nmask flag):

1. nmask.sorted.bam.
2. nmask.sorted.bam.bai.

rPGA adds up to 4 SAM tags per read:

1. SP: semicolon deliminated list of heterozygous SNP positions the read covers.
2. GT: denotes whether the read contains the reference or alternate allele(s) for the SNP(s) in SP.
3. RF: denotes the reference allele for each SNP in SP.
4. AT: denotes the alternate allele for each SNP in SP.

- For a complete list of options, please*see* Table 6.

Table 6
Options and required parameters for rPGA assign in the allele-specific assignment pipeline

Parameter	Description
-o[a]	Output directory
-v[a]	VCF file or genotype directory containing VCF files
–conflict	Flag to write bam file containing conflicting reads
–rnaedit[b]	Flag to consider RNAediting sites
-e[b]	File containing RNA editing sites
–gz	Flag denoting VCF genotype files are gzipped
-b1[c]	Haplotype 1 alignment file to personal genome (BAM)
-b2[c]	Haplotype 2 alignment file to personal genome (BAM)
–nmask[d]	Flag to n-mask variant sites (if using unphased data)

[a] Required parameter
[b] Note: if –rnaedit flag is used, a file containing RNA editing events must be provided using -e. In this case, rPGA will disregard heterozygous SNPs that overlap RNA editing sites when assigning mapped reads to haplotypes
[c] Use these if you would like to supply your own personal genomemapping alignment files. If rPGA run mapping (previous step) is used, there is no need to provide rPGA with the alignment files, and rPGA will use the alignments in HAP1/STARalign and HAP2/STARalign
[d] If –nmask flag is used, reads covering heterozygous SNPs are assigned to the reference or alternate allele. Note, –nmask option must be used for all three rPGA run steps (personalize, mapping, and alleles)

3.4 Downstream Analysis

rPGA reports the splice site usage frequency for each haplotype-specific hidden splice junction. Specifically, frequency = PJ/(PJ + OJ), where PJ is the read count of the personal specific splice junction and OJ is the combined read count of all reference splice junctions with overlapping genomic coordinates. The validation of candidate hidden splice variants can be done using a variety of techniques, most commonly RT-PCR using primers targeting flanking exon regions.

4 Notes

1. STAR is used for faster alignment. However, speed comes at the cost of required memory. Make sure you have enough memory on your computer to run STAR (32GB).
2. rPGA uses a GTF file to identify splice variants as novel or reference. The first column (chromosome/contig name) in the GTF file must match the sequence names in your FASTA reference file.

References

1. Nilsen TW, Graveley BR (2010) Expansion of the eukaryotic proteome by alternative splicing. Nature 463:457–463. Issn: 1476–4687 (Electronic) 0028–0836 (Linking)
2. Wang ET et al (2008) Alternative isoform regulation in human tissue transcriptomes. Nature 456:470–476. Issn: 1476–4687 (Electronic) 0028–0836 (Linking)
3. Hull J et al (2007) Identification of common genetic variation that modulates alternative splicing. PLoS Genet 3:e99. Issn: 1553–7404 (Electronic) 1553–7390 (Linking)
4. Kwan T et al (2008) Genome-wide analysis of transcript isoform variation in humans. Nat Genet 40:225–231. Issn: 1546–1718 (Electronic) 1061–4036 (Linking)
5. Kwan T et al (2007) Heritability of alternative splicing in the human genome. Genome Res 17:1210–1218. Issn: 1088–9051 (Print) 1088–9051 (Linking) (2007).
6. Garcia-Blanco MA, Baraniak AP, Lasda EL (2004) Alternative splicing in dis-ease and therapy. Nat Biotechnol 22:535–546. Issn: 1087–0156 (Print) 1087–0156 (Linking)
7. Ward AJ, Cooper TA (2010) The pathobiology of splicing. J Pathol 220:152–163. Issn: 1096–9896 (Electronic) 0022–3417 (Linking)
8. Lu ZX, Jiang P, Xing Y (2012) Genetic variation of pre-mRNA alternative splicing in human populations. Wiley Interdiscip Rev RNA 3:581–592. Issn: 1757–7012 (Electronic) 1757–7004 (Linking).
9. Sheth N et al (2006) Comprehensive splice-site analysis using comparative genomics. Nucleic Acids Res 34:3955–3967. Issn: 1362–4962 (Electronic) 0305–1048 (Linking)
10. Kurmangaliyev YZ, Sutormin RA, Naumenko SA, Bazykin GA, Gelfand MS (2013) Functional implications of splicing polymorphisms in the human genome. Hum Mol Genet 22:3449–3459. Issn: 1460–2083 (Electronic) 0964–6906 (Linking)
11. Krawczak M, Reiss J, Cooper DN (1992) The mutational spectrum of single base-pair substitutions in mRNA splice junctions of human genes: causes and consequences. Hum Genet 90:41–54. Issn: 0340–6717 (Print) 0340–6717 (Linking)
12. Katz Y, Wang ET, Airoldi EM, Burge CB (2010) Analysis and design of RNA sequencing experiments for identifying isoform regulation. Nat Methods 7:1009–1015. Issn: 1548–7105 (Electronic) 1548–7091 (Linking).
13. Mortazavi A, Williams BA, McCue K, Schaeffer L, Wold B (2008) Mapping and quantifying mammalian transcriptomes by RNA-Seq. Nat Methods 5:621–628. Issn: 1548–7105 (Electronic) 1548–7091 (Linking)
14. Shen S et al (2012) MATS: a Bayesian framework for flexible detection of differential alternative splicing from RNA-Seq data. Nucleic Acids Res 40:e61
15. Shen S et al (2014) rMATS: robust and flexible detection of differential alternative splicing from replicate RNA-Seq data. Proc Natl Acad Sci U S A 111:E5593–E5601. Issn: 1091–6490 (Electronic) 0027–8424 (Linking)
16. Dobin A et al (2013) STAR: ultrafast universal RNA-seq aligner. Bioinformatics 29:15–21. Issn: 1367–4811 (Electronic) 1367–4803 (Linking).
17. Trapnell C, Pachter L, Salzberg SL (2009) TopHat: discovering splice junctions with RNA-Seq. Bioinformatics 25:1105–1111. Issn: 1367–4811 (Electronic) 1367–4803 (Linking)
18. Stein S, Lu ZX, Bahrami-Samani E, Park JW, Xing Y (2015) Discover hidden splicing variations by mapping personal transcriptomes to personal genomes. Nucleic Acids Res 43:10612–10622. Issn: 1362–4962 (Electronic) 0305–1048 (Linking)

Chapter 11

Inducible Expression of Eukaryotic Circular RNAs from Plasmids

Deirdre C. Tatomer, Dongming Liang, and Jeremy E. Wilusz

Abstract

Thousands of eukaryotic protein-coding genes are noncanonically spliced to generate circular RNAs. Because they have covalently linked ends, circular RNAs are resistant to degradation by exonucleases and some accumulate to higher levels than their associated linear mRNAs. The functions of most circular RNAs are still unknown, but recent work has revealed key insights into how the pre-mRNA splicing machinery catalyzes backsplicing. Exons that circularize are often flanked by intronic repeat sequences that are complementary to one another, and backsplicing is triggered when these repeats base pair and bring the intervening splice sites into close proximity. Here, we describe how this knowledge has been translated into a simple plasmid-based method for ectopically expressing circular RNAs in eukaryotic cells. The sequence of interest is cloned into an artificial exon that is flanked by complementary intronic repeats. The plasmid is then transfected into cells, transcription is induced, and the cellular splicing machinery generates the desired circular RNA. Total RNA is isolated and the efficiency/specificity of circular RNA biogenesis is validated by Northern blot analysis. Beyond allowing overexpression of natural circular RNAs to define their functions, this approach can be used to produce designer RNA circles that are translated or bind specific cellular factors, such as microRNAs or proteins.

Key words Backsplicing, circRNA, Alternative splicing, Circularization, *Drosophila*, Laccase2, Northern blot, Transfection, Biogenesis, Noncoding RNA

1 Introduction

Most precursor mRNAs (pre-mRNAs) from eukaryotic protein-coding genes contain intronic sequences that must be removed by the spliceosome [1]. When a pre-mRNA is spliced in the canonical manner, exons are joined from 5′ to 3′ (e.g., exon 1 is joined to exon 2, which is joined to exon 3, etc.) to produce a linear mRNA that is subsequently translated (Fig. 1, top). Nevertheless, recent work has revealed thousands of pre-mRNAs that undergo backsplicing and join a splice donor to an upstream splice acceptor (e.g., the end of exon 2 is covalently joined to the beginning of exon 2), thereby generating a circular RNA (Fig. 1, bottom) [2–7]. Most

Yongsheng Shi (ed.), *mRNA Processing: Methods and Protocols*, Methods in Molecular Biology, vol. 1648,
DOI 10.1007/978-1-4939-7204-3_11,

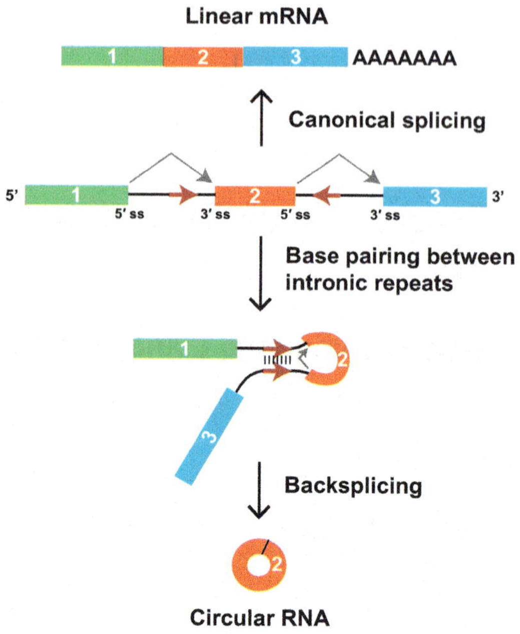

Fig. 1 A pre-mRNA can be spliced to generate a linear or circular RNA. When the splice sites (ss) are joined from 5′ to 3′, a linear mRNA is generated that is subsequently polyadenylated (*top*). Alternatively, backsplicing can be induced when inverted repeat sequences (*red arrows*) in the flanking introns base pair to one another. This brings the intervening splice sites into close proximity, facilitating the production of a circular RNA (*bottom*)

circular RNAs are rarely generated and accumulate to low levels [8], but some are expressed at levels tenfold higher than their associated linear mRNA [2, 3]. This suggests that the main function of some protein-coding genes may be to produce circular RNAs rather than proteins.

Circular RNAs are naturally resistant to degradation by exonucleases and often have half-lives greater than 24–48 h [3, 9]. In at least 2 cases (CDR1as/ciRS-7 and Sry), circular RNAs function to sequester specific microRNAs [4, 10–12], but most circular RNAs contain few microRNA binding sites [6] and likely have a different function. It is still unclear what the vast majority of circular RNAs do in cells, but artificial circular RNAs containing an internal ribosome entry site (IRES) can be translated [13–15] and circular RNAs may aid in the formation of large RNA-protein complexes [16, 17].

With the exception of the first and last exons of genes, every other exon in the genome has splice sites at its 5′ and 3′ ends and theoretically can be backspliced. However, only a small subset of possible circular RNAs are generated in cells. This is, in part, because introns are rapidly removed by the spliceosome in a

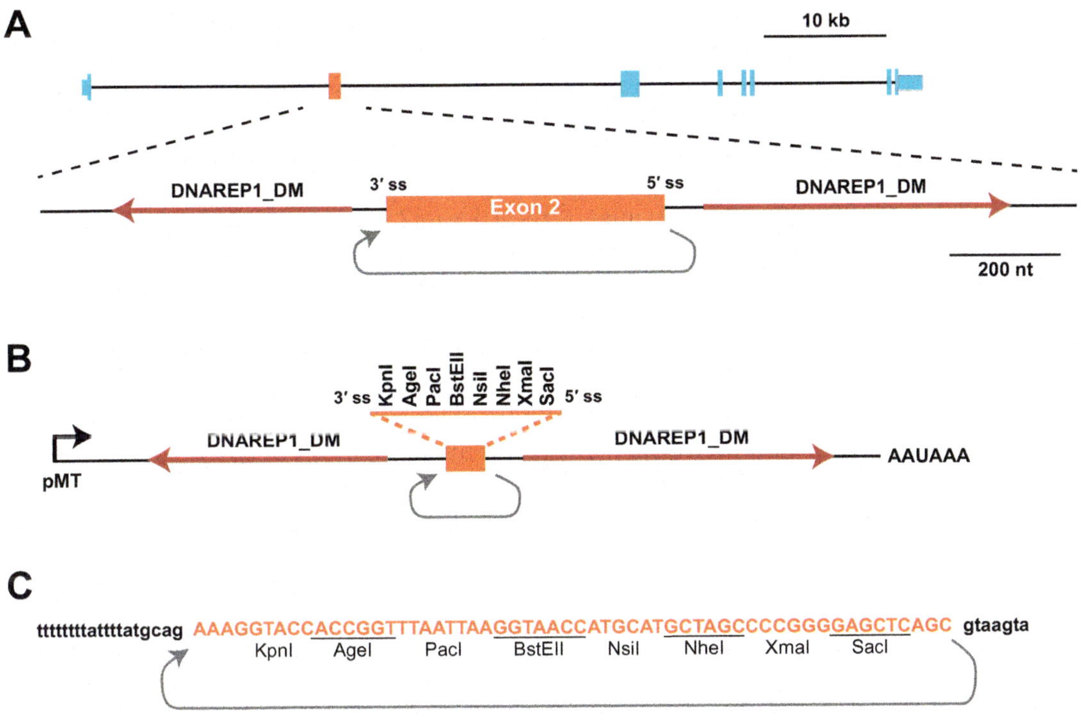

Fig. 2 Intronic repeats from the *Drosophila* Laccase2 gene facilitate circular RNA formation. (**a**) Exon/intron structure of the Laccase2 locus, highlighting a 1945-nt region that contains exon 2. Complementary *DNAREP1_DM* family transposons (*red*) flank this exon and facilitate backsplicing (*gray*). (**b**) The Hy_pMT Laccase2 MCS Exon plasmid has a multi-cloning site exon between the Laccase2 flanking introns. Transcription is driven by the metallothionein promoter (pMT) and terminated by the SV40 polyadenylation signal. The plasmid backbone contains hygromycin resistance and ampicillin resistance genes (*not shown*). Additional details were previously described [25]. (**c**) Sequence of the multi-cloning site exon (*uppercase*) and the intronic sequences that immediately flank it (*lowercase*)

co-transcriptional manner [18]. By the time a given splice donor (5′ splice site) is transcribed, all of the upstream splice acceptors (3′ splice sites) in that pre-mRNA have usually been removed, making backsplicing impossible. There are, however, a growing number of introns that are known to be slowly or posttranscriptionally spliced [19, 20]. This allows the opportunity for direct competition between canonical splicing and backsplicing [21].

At many genes, production of a circular RNA is facilitated by intronic complementary sequences (such as repetitive elements) that base pair to bring the intervening splice sites into close proximity (Fig. 1) [3, 5, 22–25]. For example, highly complementary *DNAREP1_DM* family transposons flank exon 2 of the *Drosophila* Laccase2 gene, which is backspliced to generate an abundant 490-nt circular RNA in *Drosophila* cell lines and adult heads (Fig. 2a) [25]. We showed that a 1945-nt region of the Laccase2 pre-mRNA that encompasses exon 2 and the flanking intronic repeats (zoomed region in Fig. 2a) is sufficient for the production of the Laccase2 circular RNA from plasmids in *Drosophila* and mammalian cells [25]. Disrupting base pairing between the *DNAREP1_DM* transposons eliminated Laccase2 backsplicing, while the introduction of compensatory mutations was sufficient to restore circular RNA production.

To determine if the Laccase2 flanking introns are able to support backsplicing of other exonic sequences, we replaced exon 2 with an artificial 57-nt exon composed of a multi-cloning site (MCS) to generate the Hy_pMT Laccase2 MCS Exon plasmid (Fig. 2b, c) [25]. Although the 57-nt exon was unable to be backspliced in *Drosophila* cells, the Laccase2 introns were able to efficiently circularize a variety of inserted exons that are ≥300-nt. This indicated that the flanking introns play a dominant role in circular RNA formation and can drive the intervening exon to be backspliced. Indeed, we have demonstrated that the Laccase2 introns can be used to overexpress natural circular RNAs (e.g., CDR1as/ciRS-7) as well as "designer" circular RNAs (e.g., one encoding the green fluorescent protein (GFP) open reading frame) [25]. Furthermore, the metallothionein promoter (pMT) [26] enables inducible expression of the circular RNA from these plasmids upon addition of copper (II) sulfate.

In this chapter, we describe methods for using the Hy_pMT Laccase2 MCS Exon plasmid to express circular RNAs in *Drosophila* cells. Using standard cloning techniques, the sequence of interest is inserted into the multi-cloning site present in the artificial exon. The plasmid is then transfected into DL1 or S2 cells using Effectene and expression from the metallothionein promoter is induced (Fig. 3). Total RNA is isolated from cells and Northern blot analysis is used to validate the efficiency and specificity of circular RNA biogenesis. Northern blots allow visualization of the full complement of RNAs produced from the plasmid, including undesired concatamers or linear RNAs that may be missed by other approaches like RT-PCR.

2 Materials

2.1 Generation of Circular RNA Expression Plasmid

1. Hy_pMT Laccase2 MCS Exon plasmid (*see* **Note 1**).
2. Standard reagents and equipment for performing PCR, restriction digests, ligations, and plasmid DNA purification.

2.2 Transient Transfection of Drosophila Cell Lines

1. Complete *Drosophila* Media: Schneider's *Drosophila* Medium, 10% heat-inactivated fetal bovine serum, 1% penicillin/streptomycin, and 1% L-glutamine.
2. *Drosophila* DL1 or S2 cells cultured at 25 °C.
3. 6-well tissue culture plates.

Day 1	**Seed *Drosophila* cells into 6-well plate**
Day 2	**Transfect plasmids with Effectene**
Day 3	**Add copper sulfate to cells and incubate overnight (14 h) at 25°C**
Day 4	**Isolate RNA using Trizol** **Run Northern gel (~3.5 h)** **Wash gel (2 x 15 min with $H_2$0, 2 x 15 min with 10x SSC)** **Capillary transfer overnight to membrane**
Day 5	**UV crosslink membrane** **Pre-hybridize membrane at 42°C (~1 h)** **Make / Add probe to membrane** **Hybridize overnight at 42°C**
Day 6	**Wash Northern (2 x 30 min with 2x SSC, 0.5% SDS)** **Expose to Phosphorimager screen**
Day 7	**View Phosphorimager screen**

Fig. 3 Timeline for transfecting plasmids, isolating total RNA, and analyzing circular RNA expression by Northern blot analysis. *See* the text for details

4. Circular RNA expression plasmid (*see* Subheading 2.1).
5. Effectene transfection reagent, Buffer EC, Enhancer.
6. Copper (II) sulfate: 500 mM solution in water.
7. RNA isolation reagents: TRIzol, chloroform, isopropanol, and 70% ethanol.
8. DEPC-treated water.
9. NanoDrop 2000 spectrophotometer.

2.3 Northern Blot

1. Horizontal gel apparatus (12 × 14 cm gel).
2. A 1.2% denaturing formaldehyde agarose gel: Weigh 1.8 g of agarose in a 500 mL Erlenmeyer flask and add 135 mL of ultrapure water. Microwave for ~3 min to dissolve agarose. In the meantime, warm NorthernMax 10× Denaturing Gel Buffer to 60 °C in a water bath. In a chemical fume hood, allow the agarose solution to cool for several minutes and then add 15 mL of pre-warmed NorthernMax 10× Denaturing Gel Buffer. Mix contents well and pour into gel casting tray. Use a pipet tip to remove bubbles and insert comb. Allow gel to sit for 1 h in chemical fume hood.
3. Northern gel running buffer: Add 65 mL of NorthernMax 10× Running Buffer to 585 mL of ultrapure water (*see* **Note 2**).

4. NorthernMax Formaldehyde Load Dye. Add 2–3 volumes of dye to each RNA sample.
5. Ethidium bromide (10 mg/mL) solution.
6. Power supply.
7. Variable temperature water bath.
8. Heat blocks at 65 °C and 95 °C.
9. Aluminum foil.
10. Orbital shaker with approximately 12 × 12-in. platform.
11. 10× saline-sodium citrate (SSC) buffer. Prepare 1 L for each blot.
12. Whatman 3MM chromatography paper. For each blot, cut three 12 × 14 cm sheets and one 19 × 46 cm sheet.
13. Hybond-N membrane. For each blot, cut 12 × 14 cm piece of membrane.
14. Pyrex dish (20.1 cm L × 20.1 cm W × 5.5 cm H) for washing gel.
15. Capillary transfer apparatus: Pyrex tray (34.9 cm L × 24.9 cm W × 6 cm H), glass plate (28 × 16.5 cm), plastic wrap, paper towels (Kimberly Clark KIM01700 or equivalent), and weight (e.g., catalog or textbook).
16. UV Crosslinker (Spectrolinker XL-1000).
17. ULTRAhyb-Oligo Buffer.
18. Northern probe labeling and purification: T4 polynucleotide kinase, [γ-^{32}P] ATP, oligonucleotide probe complementary to sequence of interest (*see* **Note 3**), Illustra MicroSpin G-50 columns.
19. Hybridization tubes (35 × 150 mm) and hybridization oven.
20. Northern wash buffer: 2× SSC, 0.5% sodium dodecyl sulfate (SDS).
21. Plastic wrap.
22. Phosphorimager screen and scanner (Typhoon FLA 9500).

3 Methods

3.1 Generation of Circular RNA Expression Plasmid

1. Use PCR to amplify region of interest flanked by restriction sites for cloning (*see* **Note 4**). The Hy_pMT Laccase2 MCS Exon plasmid has been successfully used to express circular RNAs from 300 to 1500-nt in length, and it should be noted that exons less than 300-nt generally fail to be backspliced [25].
2. Use standard cloning techniques to cut the PCR product and Hy_pMT Laccase2 MCS Exon plasmid with appropriate

restriction enzymes. Purify the digested DNA fragments from an agarose gel, ligate the insert and vector together, and transform into chemically competent *E. coli* cells. Plate on ampicillin-treated LB agar.

3. Screen colonies for proper insertion of the region of interest by PCR, restriction enzyme digestion, and/or DNA sequencing. Prepare plasmid DNA using a commercially available kit.

3.2 Transient Transfection of Drosophila Cell Lines

1. Grow DL1 or S2 cells in complete *Drosophila* media until confluent (*see* **Note 5**). On the day before transfection (Day 1, Fig. 3), seed each well of a 6-well plate with 2 × 10^6 cells in 2 mL of complete media. Distribute cells uniformly by rocking the plate in side-to-side motions. Incubate cells at 25 °C.
2. On the day of transfection (Day 2, Fig. 3), prepare transfection reagent mixtures by combining 2 μg of plasmid DNA, 200 μL Buffer EC, and 16 μL Enhancer in sterile 1.7 mL tubes. A mock transfection sample containing Buffer EC and Enhancer, but no plasmid DNA should also be prepared. Vortex mixtures for 1 s and incubate at room temperature for 5 min.
3. Add 8 μL Effectene reagent to each sample, vortex for 5 s, and incubate at room temperature for 10 min.
4. Remove 6-well plate from incubator and dispense all of the transfection mixture dropwise into the appropriate well. Mix gently by rocking to distribute the mixture across the cells.
5. Incubate cells at 25 °C for 24–36 h.
6. Add 2 μL copper (II) sulfate to the media (final concentration of 500 μM) to induce transcription from the plasmid's metallothionein promoter.
7. After 14 h, harvest cells by removing the media and adding 1 mL of TRIzol to each well. Isolate total RNA according to the manufacturer's protocol. Resuspend RNA in 50 μL DEPC-treated water. Determine RNA concentration using a NanoDrop spectrophotometer.

3.3 Northern Blot to Detect Circular RNA Expression

1. In a chemical fume hood, prepare a horizontal gel apparatus containing a 1.2% agarose-formaldehyde gel and Northern gel running buffer. Cover gel apparatus with aluminum foil to protect from light.
2. Prepare RNA samples by adding 10–20 μg of RNA to 2–3 volumes of NorthernMax Formaldehyde Loading Dye (*see* **Note 6**). Incubate samples at 65 °C for 15 min to denature RNA secondary structures. Place samples on ice for 5 min, briefly spin, and add 0.5 μL ethidium bromide to each tube (*see* **Note 7**).

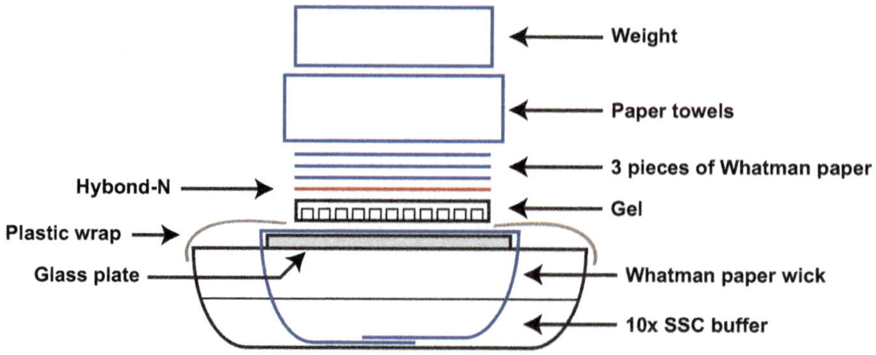

Fig. 4 Capillary transfer apparatus used to transfer RNA from the formaldehyde agarose gel to the membrane

3. Load gel in chemical fume hood and run at 95 V for 3–3.5 h (*see* **Note 8**). Cover gel apparatus with aluminum foil to protect from light.
4. Disassemble gel apparatus and place gel/casting tray on UV box to visualize ribosomal RNAs and verify RNA quality. The ribosomal RNA bands should be prominent with no smearing.
5. Remove gel from casting tray and place in a clean Pyrex dish. Add ~250 mL ultrapure water and shake gently on an orbital shaker for 15 min.
6. During this wash, begin to set up the capillary transfer apparatus (Fig. 4). Use ~600 mL of 10× SSC to fill a Pyrex tray. Fold a 19 × 46 cm sheet of Whatman 3MM chromatography paper in a C-fold around a glass plate and place perpendicularly on the top of the Pyrex tray. Ensure the Whatman paper wick ends are immersed in 10× SSC.
7. Decant water from Pyrex dish containing the gel and add ~250 mL of fresh ultrapure water. Place on an orbital shaker for 15 min. Decant and perform two 15 min washes with ~200 mL 10× SSC. Place on an orbital shaker for 15 min each time. In total, there is 1 h of gel washes (2 × 15 min in water and 2 × 15 min in 10× SSC).
8. Set up capillary transfer apparatus (Fig. 4). Smooth the Whatman paper wick surface by using a pipet to roll out air bubbles. Pipet 10 mL 10× SSC onto the wick and place the gel upside down in the middle of the wick. Placing the gel upside down allows a shorter distance for the RNA to travel to the membrane. It is important to avoid air bubbles between all the layers of the transfer apparatus, so use a pipet to roll out air bubbles between the wick and gel (*see* **Note 9**). Use the Pyrex dish that had contained the gel to wet a 12 × 14 cm sheet of

Hybond-N membrane in 10× SSC. This membrane is the same size as the gel and should be placed directly above the gel. Do not attempt to move the membrane once in contact with the gel. Wet three pieces of 12 × 14 cm Whatman 3MM chromatography paper with 10× SSC and place sequentially on the top of the membrane. The remaining 10× SSC buffer from the Pyrex dish can then be added to the ~600 mL of 10× SSC used for the transfer (from **step 6**). Use a piece of plastic wrap to cover the entire transfer apparatus to minimize evaporation, and then cut around the gel area to make a window. Place a stack of paper towels (approximately 7 cm) on the top of the last 12 × 14 cm Whatman paper. Place a weight (e.g., catalog or textbook) on top. Allow transfer to proceed overnight.

9. Disassemble transfer apparatus by sequentially removing each layer from the top. Turn the membrane over and place on a paper towel so that the side with the RNA transferred to it is face up. Cut the top left hand corner of the membrane to mark orientation. Examine the gel under UV light to verify that ribosomal RNAs have efficiently transferred.
10. Place the membrane in UV Crosslinker. Use "Optimal Crosslink" button, which delivers an energy dosage of 120 mJ per unit area.
11. Preheat ULTRAhyb-Oligo buffer to 50 °C in a water bath. Swirl the bottle to fully dissolve any precipitated material.
12. Cut the membrane to remove unused lanes (*see* **Note 10**). To simplify cutting, a handheld 254-nm UV light source can be used for UV shadowing to directly visualize the ribosomal RNAs.
13. Place the membrane in a hybridization tube so that the side with the RNA transferred to it faces the interior of the tube. The membrane should be rolled in such a way as to minimize overlap in the tube. Add 10 mL of preheated ULTRAhyb-Oligo buffer. Rotate the membrane in a hybridization oven at 42 °C for at least 45 min.
14. Prepare radiolabeled probe against circular RNA by labeling DNA oligo with T4 polynucleotide kinase (PNK) and [γ-^{32}P] ATP (*see* **Note 11**). Typical labeling reactions contain 11.5 μL ultrapure water, 1.5 μL DNA oligo (10 μM), 2 μL 10× PNK Buffer, 4 μL [γ-^{32}P] ATP, and 1 μL T4 polynucleotide kinase. Incubate at 37 °C for 45–60 min. Stop the reaction by incubating at 95 °C for 5 min. Add 30 μL ultrapure water to bring total volume to 50 μL. Remove unincorporated [γ-^{32}P] ATP using Illustra MicroSpin G-50 columns according to the manufacturer's instructions.
15. Add 15 μL of probe to hybridization tube. Rotate and hybridize membrane overnight at 42 °C.

16. On the next day, preheat Northern wash buffer to 42 °C in a water bath (*see* **Note 12**).
17. Discard probe solution from hybridization tube into appropriate radioactive waste container. Immediately add ~30 mL of pre-warmed Northern wash buffer to hybridization tube. Rotate for 30 min at 42 °C to wash membrane. Discard wash buffer and add fresh Northern wash buffer. Rotate for 30 min at 42 °C.
18. Remove membrane from last wash and wrap in plastic wrap. Expose to a Phosphorimager screen. Signal can usually be observed within 2 h, but exposures of 24–72 h are needed for high-quality images. Scan the screen using a Typhoon FLA 9500.

4 Notes

1. Hy_pMT Laccase2 MCS Exon plasmid is available from the Wilusz lab or Addgene (https://www.addgene.org/69884/). This chapter is focused on expression of circular RNAs in *Drosophila* cells, but methods for expressing circular RNAs in mammalian cells have been described [25]. pcDNA3.1(+) Laccase2 MCS Exon Vector (https://www.addgene.org/69893/) or pcDNA3.1(+) ZKSCAN1 MCS Exon Vector (https://www.addgene.org/69901/) should be used for mammalian cells [25]. The method is analogous to the one described here, except that Lipofectamine 2000 is used for the transfections and a constitutive CMV promoter is present in the plasmids.
2. The running buffer contains MOPS and should be protected from light.
3. We typically use ~20 nucleotide oligonucleotides with a T_m of 60 °C that were purified by standard desalting. To detect both linear and circular RNAs generated from the Hy_pMT Laccase2 MCS Exon plasmid, we recommend a primer complementary to the XmaI and SacI restriction sites (5′-GCTGAGCTCCCCGGG).
4. We typically use KpnI and XmaI for the insertion into the Hy_pMT Laccase2 MCS Exon plasmid, but other restriction sites present in the multi-cloning site exon can also be used (Fig. 2c).
5. Basic protocols for *Drosophila* cell culture are available on the DRSC Functional Genomics Resources (DRSC-FGR) website (http://fgr.hms.harvard.edu/fly-cell-culture) [27].
6. Ensure that the RNA sample is sufficiently concentrated to fit in the well. 30 μL can fit in each well when the 20-well comb (1.5 mm thickness) is used. 45 μL can fit when the 16-well comb (1.5 mm thickness) is used.

7. Addition of ethidium bromide is not mandatory, as it simply enables visualization of the ribosomal RNAs.
8. Stop electrophoresis when the bromophenol blue dye front has migrated to approximately 1 cm from the bottom of the gel.
9. A pipet should be similarly used to remove air bubbles after the Hybond-N membrane and each piece of Whatman paper have been added to the transfer apparatus.
10. We generally leave at least one or two lanes unused, so that the membrane will not overlap with itself when placed in the hybrization tube.
11. To determine the amounts of linear vs. circular RNAs produced from the plasmid, distinct oligonucleotide probes should be tested. A probe complementary to the middle of the exon should detect all RNAs (linear or circular) generated from the plasmid, whereas a probe complementary to the backspliced junction should only detect circular RNAs. Resistance to RNase R treatment can also be used to verify circular RNA formation [25].
12. It is important that the Northern wash buffer is preheated to avoid high background signal on the membrane.

Acknowledgments

We thank Sara Cherry for introducing our lab to *Drosophila* tissue culture methods. This work was supported by NIH grants R00-GM104166 and R35-GM119735. J.E.W. is a Rita Allen Foundation Scholar.

References

1. Fu XD, Ares M Jr (2014) Context-dependent control of alternative splicing by RNA-binding proteins. Nat Rev Genet 15(10):689–701. doi:10.1038/nrg3778
2. Salzman J, Gawad C, Wang PL, Lacayo N, Brown PO (2012) Circular RNAs are the predominant transcript isoform from hundreds of human genes in diverse cell types. PLoS One 7(2):e30733. doi:10.1371/journal.pone.0030733
3. Jeck WR, Sorrentino JA, Wang K, Slevin MK, Burd CE, Liu J, Marzluff WF, Sharpless NE (2013) Circular RNAs are abundant, conserved, and associated with ALU repeats. RNA 19(2):141–157. doi:10.1261/rna.035667.112
4. Memczak S, Jens M, Elefsinioti A, Torti F, Krueger J, Rybak A, Maier L, Mackowiak SD, Gregersen LH, Munschauer M, Loewer A, Ziebold U, Landthaler M, Kocks C, le Noble F, Rajewsky N (2013) Circular RNAs are a large class of animal RNAs with regulatory potency. Nature 495(7441):333–338. doi:10.1038/nature11928
5. Zhang XO, Wang HB, Zhang Y, Lu X, Chen LL, Yang L (2014) Complementary sequence-mediated exon circularization. Cell 159(1):134–147. doi:10.1016/j.cell.2014.09.001
6. Guo JU, Agarwal V, Guo H, Bartel DP (2014) Expanded identification and characterization of mammalian circular RNAs. Genome Biol 15(7):409. doi:10.1186/s13059-014-0409-z
7. Wilusz JE (2016) Circular RNAs: unexpected outputs of many protein-coding genes. RNA Biol. doi:10.1080/15476286.2016.1227905
8. Zhang Y, Xue W, Li X, Zhang J, Chen S, Zhang JL, Yang L, Chen LL (2016) The biogenesis of

nascent circular RNAs. Cell Rep 15(3):611–624. doi:10.1016/j.celrep.2016.03.058

9. Enuka Y, Lauriola M, Feldman ME, Sas-Chen A, Ulitsky I, Yarden Y (2016) Circular RNAs are long-lived and display only minimal early alterations in response to a growth factor. Nucleic Acids Res 44(3):1370–1383. doi:10.1093/nar/gkv1367
10. Hansen TB, Jensen TI, Clausen BH, Bramsen JB, Finsen B, Damgaard CK, Kjems J (2013) Natural RNA circles function as efficient microRNA sponges. Nature 495(7441):384–388. doi:10.1038/nature11993
11. Wilusz JE, Sharp PA (2013) Molecular biology. A circuitous route to noncoding RNA. Science 340(6131):440–441. doi:10.1126/science.1238522
12. Westholm JO, Miura P, Olson S, Shenker S, Joseph B, Sanfilippo P, Celniker SE, Graveley BR, Lai EC (2014) Genome-wide analysis of *Drosophila* circular RNAs reveals their structural and sequence properties and age-dependent neural accumulation. Cell Rep 9(5):1966–1980. doi:10.1016/j.celrep.2014.10.062
13. Chen CY, Sarnow P (1995) Initiation of protein synthesis by the eukaryotic translational apparatus on circular RNAs. Science 268(5209):415–417
14. Wang Y, Wang Z (2015) Efficient backsplicing produces translatable circular mRNAs. RNA 21(2):172–179. doi:10.1261/rna.048272.114
15. Abe N, Matsumoto K, Nishihara M, Nakano Y, Shibata A, Maruyama H, Shuto S, Matsuda A, Yoshida M, Ito Y, Abe H (2015) Rolling circle translation of circular RNA in living human cells. Sci Rep 5:16435. doi:10.1038/srep16435
16. Rybak-Wolf A, Stottmeister C, Glazar P, Jens M, Pino N, Giusti S, Hanan M, Behm M, Bartok O, Ashwal-Fluss R, Herzog M, Schreyer L, Papavasileiou P, Ivanov A, Ohman M, Refojo D, Kadener S, Rajewsky N (2015) Circular RNAs in the mammalian brain are highly abundant, conserved, and dynamically expressed. Mol Cell 58(5):870–885. doi:10.1016/j.molcel.2015.03.027
17. You X, Vlatkovic I, Babic A, Will T, Epstein I, Tushev G, Akbalik G, Wang M, Glock C, Quedenau C, Wang X, Hou J, Liu H, Sun W, Sambandan S, Chen T, Schuman EM, Chen W (2015) Neural circular RNAs are derived from synaptic genes and regulated by development and plasticity. Nat Neurosci 18(4):603–610. doi:10.1038/nn.3975
18. Carrillo Oesterreich F, Herzel L, Straube K, Hujer K, Howard J, Neugebauer KM (2016) Splicing of nascent RNA coincides with intron exit from RNA polymerase II. Cell 165(2):372–381. doi:10.1016/j.cell.2016.02.045
19. Boutz PL, Bhutkar A, Sharp PA (2015) Detained introns are a novel, widespread class of post-transcriptionally spliced introns. Genes Dev 29(1):63–80. doi:10.1101/gad.247361.114
20. Vargas DY, Shah K, Batish M, Levandoski M, Sinha S, Marras SA, Schedl P, Tyagi S (2011) Single-molecule imaging of transcriptionally coupled and uncoupled splicing. Cell 147(5):1054–1065. doi:10.1016/j.cell.2011.10.024
21. Ashwal-Fluss R, Meyer M, Pamudurti NR, Ivanov A, Bartok O, Hanan M, Evantal N, Memczak S, Rajewsky N, Kadener S (2014) circRNA biogenesis competes with pre-mRNA splicing. Mol Cell 56(1):55–66. doi:10.1016/j.molcel.2014.08.019
22. Capel B, Swain A, Nicolis S, Hacker A, Walter M, Koopman P, Goodfellow P, Lovell-Badge R (1993) Circular transcripts of the testis-determining gene Sry in adult mouse testis. Cell 73(5):1019–1030
23. Pasman Z, Been MD, Garcia-Blanco MA (1996) Exon circularization in mammalian nuclear extracts. RNA 2(6):603–610
24. Liang D, Wilusz JE (2014) Short intronic repeat sequences facilitate circular RNA production. Genes Dev 28(20):2233–2247. doi:10.1101/gad.251926.114
25. Kramer MC, Liang D, Tatomer DC, Gold B, March ZM, Cherry S, Wilusz JE (2015) Combinatorial control of *Drosophila* circular RNA expression by intronic repeats, hnRNPs, and SR proteins. Genes Dev 29(20):2168–2182. doi:10.1101/gad.270421.115
26. Bunch TA, Grinblat Y, Goldstein LS (1988) Characterization and use of the *Drosophila* metallothionein promoter in cultured *Drosophila melanogaster* cells. Nucleic Acids Res 16(3):1043–1061
27. Flockhart IT, Booker M, Hu Y, McElvany B, Gilly Q, Mathey-Prevot B, Perrimon N, Mohr SE (2012) FlyRNAi.Org–the database of the *Drosophila* RNAi screening center: 2012 update. Nucleic Acids Res 40(Database issue):D715–D719. doi:10.1093/nar/gkr953

Chapter 12

Methods for Extraction of RNA, Proteins, or Protein Complexes from Subcellular Compartments of Eukaryotic Cells

Kyu-Hyeon Yeom and Andrey Damianov

Abstract

Eukaryotic organelles or subcellular compartments can be selectively isolated based on their physical density and their stability in the presence of nonionic detergents. This chapter describes a protocol for the preparation of cytoplasm, nucleoplasm, and chromatin, in addition to isolation of RNA and proteins from these fractions. Proteins and protein complexes stably associated with chromatin and other high molecular weight nuclear components can be extracted under non-denaturing conditions by enzymatic digestion of RNA and DNA. The chapter also includes a detailed extraction protocol from highly purified nuclei.

Key words Subcellular fractionation, Cytoplasm, Nucleoplasm, Chromatin, High molecular weight nuclear extract

1 Introduction

In eukaryotes, subcellular compartmentalization provides distinguishable (or unique) locations for each cellular process. Beyond the fact that transcription and translation occur in the nucleus and the cytoplasm respectively, the destination of each macromolecule, being an RNA or a protein, can be altered to where they are currently needed. For example, the shuttling of many proteins between the nucleus and the cytoplasm is controlled by posttranslational modification, which is also a way to control their activity [1, 2]. Small RNAs that mature and function in the cytoplasm can be imported back to the nucleus where they play additional roles [3, 4]. Therefore, it is essential to know where each molecule of interest is localized in the cell. Even though this is typically assayed by microscopy methods, it is useful to biochemically fractionate subcellular compartments for functional in-depth analyses.

Yongsheng Shi (ed.), *mRNA Processing: Methods and Protocols*, Methods in Molecular Biology, vol. 1648,
DOI 10.1007/978-1-4939-7204-3_12, © Springer Science+Business Media LLC 2017

Subcellular fractionation methods have a history of more than a century [5]. Since the methods inception it has been significantly improved by efforts of scientists in different fields. Opening the cell is the first step and can be achieved in several ways, including mechanical force (homogenization in douncers or grinders, ultrasound sonication) or mild nonionic detergent (Digitonin, Triton X-100, Tween 20, and Igepal CA-630). The key point of this step is to break the cell membrane without disturbing the integrity of other organelles. The resulting homogenate then is applied to differential centrifugation followed by density-gradient centrifugation. The principle behind this is that cell components and organelles are separated by their size and density.

In the nucleus, transcription is largely coupled to pre-mRNA processing events, thus nascent RNA, associated with chromatin via the RNA polymerase II complex, is also bound by splicing and 3′ end processing factors. Separating the precipitable chromatin-associated material from the soluble nucleoplasm typically requires urea or stringent salt. However, some protein complexes can be destabilized under these conditions. An alternative method for releasing protein complexes from the nuclear pellet is an enzymatic extraction by nucleases that digest RNA, DNA, or both. This approach is useful in both mechanistic and functional studies of gene expression and instrumental in understanding the regulatory control over information processing.

This chapter introduces a method to fractionate cells into cytoplasmic, soluble nucleoplasmic, and chromatin fractions for RNA and protein extraction in small, micro-centrifuge tube scale [6, 7]. Later in this chapter, we describe a detailed method about how to prepare protein complexes from soluble nucleoplasm and high molecular weight nuclear material [8].

Note that the single-step purification described here may not be sufficient to obtain pure fractions. Contamination with a small amount of material from other compartments may be expected.

2 Subcellular Fractionation for Protein and RNA Extraction

We describe a fractionation method for extracting proteins and RNA from three subcellular compartments: cytoplasm, nucleoplasm, and chromatin. In brief, cells are allowed to swell in hypotonic buffer, and treated with nonionic detergent. Under these conditions the cellular membrane is lysed, while the nuclei remain intact. The nuclei are then separated from the cytoplasmic material by pelleting through a 24% sucrose solution. The nuclei are incubated in a buffer containing glycerol, urea and a higher percentage of detergent to lyse the nuclear membrane. The chromatin fraction is spun down by centrifugation, while the nucleoplasmic material remains in the supernatant. An example of this fractionation is shown in Fig. 1.

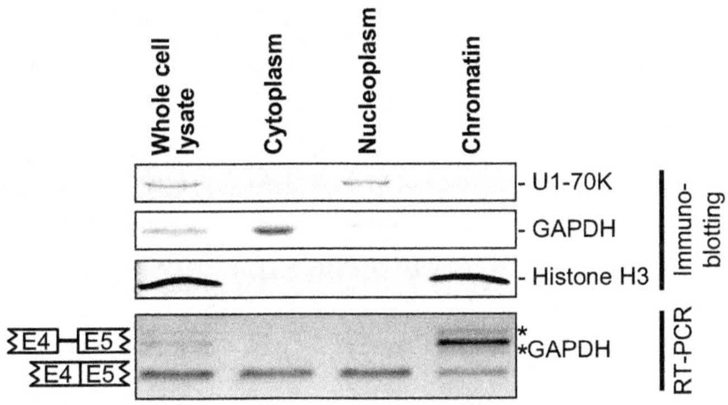

Fig. 1 Subcellular fractionation of N2a neuroblastoma cells. Protein and RNA samples were prepared from cytoplasm, nucleoplasm, and chromatin fractions of N2a cells. The relative amounts of U1-70 K, GAPDH, and Histone H3 proteins in these fractions were detected by immunoblotting (*top*). The distribution of GAPDH mRNA and pre-mRNA was tested using RT-PCR (*bottom*). Asterisks indicate nonspecific bands

2.1 Materials

Make a large volume of solution to minimize variations between batches. All buffers should be prepared using RNase-free components. Detergent and glycerol homogeneity in these solutions is achieved by rotating overnight. DTT, PMSF, Spermidine, and Spermine are added just before use.

Protease and phosphatase inhibitors are added before use at working concentration. Store all reagents at 4 °C unless otherwise indicated.

2.1.1 Solutions

1. PBS/EDTA: 1× PBS and 1 mM EDTA. Add 0.5 mM EDTA [pH 8.0] to 1× PBS as a final concentration of 1 mM.
2. 0.025% Trypsin in 1× PBS: Dilute 2.5% Trypsin solution in 1× PBS to a final concentration of 0.025%.
3. Hypotonic buffer: 10 mM Tris–HCl [pH 7.5], 15 mM KCl, 1 mM EDTA [pH 8.0], 0.15 mM Spermine, 0.5 mM Spermidine, 0.5 mM DTT, and 1× Protease inhibitor.
4. 2× Detergent Cytoplasmic Membrane Lysis buffer (2× DCML buffer): 2× detergent (e.g., Igepal CA-60), 10 mM Tris–HCl [pH 7.5], 15 mM KCl, 1 mM EDTA [pH 8.0], 0.15 mM Spermine, 0.5 mM Spermidine, 0.5 mM DTT, and 1× Protease inhibitor (*see* **Note 7**).
5. 24% Sucrose solution: 10 mM Tris–HCl [pH 7.5], 15 mM KCl, 1 mM EDTA [pH 8.0], 24% sucrose (w/v), 0.15 mM Spermine, 0.5 mM Spermidine, 0.5 mM DTT, and 1× Protease inhibitor. Sucrose buffer gets easily spoiled, so make it fresh or use within a week.

6. Glycerol buffer: 20 mM Tris–HCl [pH 8.0], 75 mM NaCl, 0.5 mM EDTA [pH 8.0], 50% glycerol (v/v), 0.85 mM DTT, and 0.125 mM PMSF.
7. Cold nuclei lysis buffer: 20 mM HEPES [pH 7.6], 1 mM DTT, 7.5 mM $MgCl_2$, 0.2 mM EDTA [pH 8.0], 0.3 M NaCl, 1 M Urea, 1% Igepal CA-630, 0.3 mM Spermine, 1.0 mM Spermidine, 1× Protease inhibitor.
8. 5% SDS lysis buffer: 5% SDS, 30% glycerol (v/v) and 150 mM Tris–HCl [pH 8.0]. Store at RT.

2.1.2 Consumables

1. Low-adhesion 1.5 mL or 1.7 mL tube.
2. Large-orifice 200 μL tips.
3. RNase-free Recombinant DNase I, 5 units/μL.
4. Trizol and Trizol LS solutions.
5. Benzonase nuclease, 250 units/μL.

2.1.3 Antibodies for immunoblotting

1. Anti-U1-70 K.
2. Anti-GAPDH.
3. Anti-Histone H3.

2.1.4 Primers for RT-PCR Amplification of mouse GAPDH

1. Reverse transcription primer: CACTGTCACACCAGAGAC AAGC.
2. PCR amplification primers:

   Forward: AACTTTGGCATTGTGGAAGG.

   Reverse: CACATTGGGGGTAGGAACAC.

2.2 Methods

This protocol is optimized for 5–10 million cells per 1.5 mL conical tube. Carry out all procedures on ice and use ice-cold buffers unless otherwise specified.

1. Prepare ~90% confluent cells on culture plate at the time of experiment (*see* **Note 1**).
2. Wash the plate twice with PBS/EDTA. Trypsinize cells by adding 1 mL of 0.025% Trypsin and keep for 3–5 min in a tissue culture incubator at 37 °C until the cells detach from the plate. Transfer the trypsinized cells to 50 mL conical tube (~ 1 mL) and pipette gently three times in the tube. Collect residual cells on the plate with 3 mL of FBS containing growth medium. Centrifuge at 100 × *g* for 5 min at 4 °C. Remove the medium by vacuum aspiration and loosen the remaining pellet by gentle flicking (*see* **Note 2**).
3. Resuspend the pelleted cells in 1 mL of PBS/EDTA and determine the cell concentration (*see* **Note 3**). Gently transfer 5–10

million cells to a 1.5 mL tube (*see* **Note 4**) and add PBS/EDTA up to 1.5 mL. Centrifuge at 100 × *g* for 10 min at 4 °C (*see* **Note 5**).

4. Remove as much PBS/EDTA supernatant as possible. Add 100 μL of hypotonic buffer to the cell pellet and flick several times. Pelleted cells should be completely resuspended. Keep on ice for 3 min (*see* **Note 6**).
5. Add 100 μL of 2× DCML buffer and gently mix by flicking 3–5 times immediately. Keep on ice for 1–4 min (*see* **Note 7**).
6. Overlay the entire lysate (~200 μL) on top of 500 μL of 24% sucrose solution in a low adhesion tube using a large-orifice 200 μL tip. Centrifuge at 6,000 × *g* for 10 min at 4 °C (*see* **Note 8**). Transfer the cytoplasmic fraction above the sucrose solution to a fresh tube. Remove the sucrose solution and save the pelleted nuclei (*see* **Note 9**).
7. [Optional] Transfer the nuclear pellet to a new 1.5 mL tube with ~100 μL of PBS/EDTA using a large-orifice 200 μL tip (*see* **Note 10**).
8. Wash the nuclei with 0.8–1 mL of PBS/EDTA. Invert the tube gently to disperse the pellet (*see* **Note 11**). Centrifuge at 6,000 × *g* for 2 min at 4 °C. Remove the supernatant as much as possible (*see* **Note 12**).
9. Add 100 μL of glycerol buffer to the nuclei pellet and quickly spin for 1 s in a tabletop centrifuge (*see* **Note 13**). Flick tube gently to disperse the nuclei (*see* **Note 14**). When the entire material is resuspended, add 100 μL of cold nuclei lysis buffer and mix thoroughly by vortexing at medium speed for 2 s, twice. Transfer 50% of the nuclear lysate to a fresh tube, and incubate both on ice for 2 min (*see* **Note 15**). Centrifuge at 6,000 × *g* for 2 min at 4 °C. Combine the nucleoplasmic supernatant from both aliquots in a fresh tube (*see* **Note 16**).
10. Rinse the pelleted chromatin by adding 500 μL of PBS/EDTA without disturbing the pellet and remove the buffer immediately using a 1,000 μL tip. Add 1 mL of PBS/EDTA and invert three times. Centrifuge at 6,000 × *g* for 2 min at 4 °C. Remove the solution completely (*see* **Note 17**). Extract proteins or RNA from each chromatin aliquot immediately (*see* **Note 18**).
11. Take 5% from each fraction for immunoblotting analysis to evaluate fractionation purity (*see* **Notes 19** and **20**).
12. Extract RNA from the cytoplasmic and nucleoplasmic fractions using Trizol LS followed by DNase I treatment for RT-PCR to evaluate fractionation purity (*see* **Note 21**).

3 Preparation of Protein Extracts from Soluble and High Molecular Weight Nuclear Fractions

This protocol is optimized for extraction of nuclear RNA and DNA binding proteins under mild, non-denaturing conditions, and is useful for studying their interactions with other proteins. In brief, highly purified nuclei are prepared by ultracentrifugation through a hypotonic, high-sucrose cushion [9]. The pelleted nuclei are lysed in the presence of mild salt and nonionic detergent, and the soluble nucleoplasm is then separated from the high molecular weight insoluble material. Note that these nuclear fractions are separated in the absence of urea and thus they differ from the nucleoplasm and chromatin described above. Free proteins and protein complexes from these two nuclear fractions are subsequently extracted by incubation with Benzonase, which digests both RNA and DNA [8]. An example demonstrating this fractionation and extraction method is shown in Fig. 2.

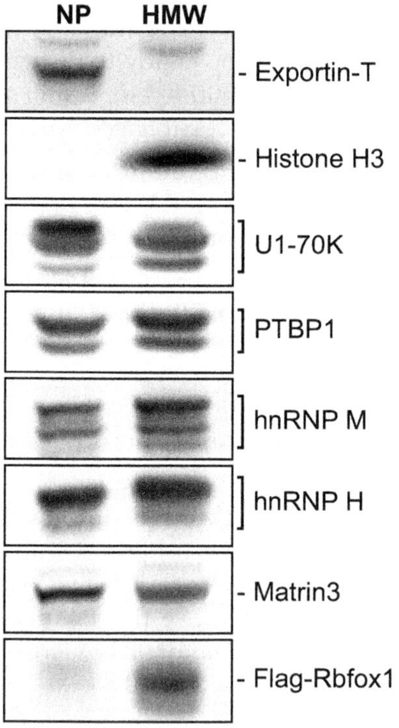

Fig. 2 Protein extraction from nucleoplasmic and high molecular weight fractions of HEK293 cells expressing Flag-tagged Rbfox1 protein. RNA binding proteins extracted from the nucleoplasm (NP) and the high molecular weight (HMW) nuclear material were detected by immunoblotting as indicated on the *right*. Exportin-T and Histone H3 were used as markers for nucleoplasm and chromatin, respectively

3.1 Materials

3.1.1 Solutions

1. Wash buffer, 10 mM HEPES-KOH [pH 7.6], 15 mM KCl, 1 mM EDTA, 0.15 mM Spermine, and 0.5 mM Spermidine.
2. Homogenization buffer, 10 mM HEPES-KOH [pH 7.6], 15 mM KCl, 1 mM EDTA, 2.2 M sucrose, 5% Glycerol, 0.15 mM Spermine, and 0.5 mM Spermidine.
3. Cushion buffer, 10 mM HEPES-KOH [pH 7.6], 15 mM KCl, 1 mM EDTA, 2.0 M sucrose, 10% Glycerol, 0.15 mM Spermine, and 0.5 mM Spermidine.
4. HBSS with calcium and magnesium, without Phenol Red.
5. PBS, 137 mM NaCl, 3 mM KCl, 80 mM Na_2HPO_4, and 20 mM KH_2PO_4.

3.1.2 Consumables

1. Benzonase nuclease, 250 units/μL.
2. Centrifugation tubes 14 × 89 mm, ultra-clear.
3. Centrifugation tubes 25 × 89 mm, ultra-clear.

3.1.3 Antibodies for immunoblotting

1. Anti-Exportin-T.
2. Anti-U1-70 K [10].
3. Anti-PTBP1 [11].
4. Anti-hnRNP M.
5. Anti-hnRNP H [12].
6. Anti-Matrin3.
7. Anti-Flag.
8. Anti-Histone H3.

3.2 Methods

1. Biological samples.
   (a) *Brain samples.* Rinse the freshly dissected tissue with HBSS buffer and mince with a clean blade on ice.
   (b) *Suspension cell culture.* Pellet down by low-RPM centrifugation, wash the cells by resuspending in 20–50 volumes of 1× PBS, then pellet down again, and discard the supernatant.
   (c) *Adherent cell culture.* Grow the cells in monolayer, remove the medium, add PBS buffer, and scrape from the plate by mechanical force (trypsinization is not recommended), then proceed as with suspension cultures.
2. Determine the volume of the cell pellet or minced tissue. Transfer the sample into a Wheaton Potter-Elvehjem style tissue grinder with a motor-driven pestle with nine volumes of ice-cold homogenization buffer (*see* **Notes 22** and **23**).
3. Homogenize on ice. Typically one full down-and-up stroke is sufficient to homogenize most samples. Having a second person to assist by operating the motor power control is highly

advisable. Both persons must wear eye protection during this step (*see* **Note 24**).

4. Transfer the homogenate to an ultracentrifugation tube containing 3.5 mL (for Beckman SW41 rotor) or 10 mL (for Beckman SW28 rotor) of cushion buffer. The homogenate should overlay the cushion without mixing with it (*see* **Note 25**).
5. Centrifuge for 1 h at 4 °C, 135,000 × *g* (28,100 RPM for SW41 and 27,000 RPM for SW28). The nuclei form a whitish pellet at the bottom, often nearly invisible. The other cellular components remain in the homogenate or at the interphase with the cushion. A thick layer of lipids may cover the top.
6. Remove the lipid layer with a spatula. Remove the homogenate and about one third of the cushion by pipetting. Mark the current level on the side wall. Pour out the rest of the cushion by inverting the tube. While keeping the tube upside down, carefully make a horizontal cut 1.5–2 cm above the bottom, but below the mark. Invert the bottom part containing the nuclear pellet, and place it on ice.
7. Using a large-orifice 200 μL or 1,000 μL tip, resuspend the nuclei in 1,000–1,500 μL of wash buffer, then pellet the nuclei at 1,000 × *g* for 5 min at 4 °C using a swinging bucket rotor.
8. Using a large-orifice 200 μL or 1,000 μL tip, resuspend the nuclei in 10 volumes of ice-cold lysis buffer. Avoid creating a lot of foam (*see* **Note 26**).
9. Transfer the nuclear lysate into a fresh tube. To allow complete lysis, keep the tube for 5 min on ice mixing gently about once per minute, or place on a rotator at 4 °C. Centrifuge for 5 min at 20,000 × *g*, 4 °C. The pelleted high molecular weight material should be less than 1/10 of the total volume. Transfer the supernatant, which contains the soluble fraction, or the nucleoplasm, into a fresh tube and measure its volume. Add an equal amount of lysis buffer to the high molecular weight pellet.
10. Add Benzonase nuclease to each fraction, to a final concentration of 2–7 unit/μL (*see* **Notes 27** and **28**). Detach the high molecular weight pellet from the bottom of the tube using a micro tip. Place both tubes on a rotator at 25 °C. The Benzonase will slowly resuspend the pellet. To speed up this process, try breaking the pellet into smaller pieces by pipetting with a large-orifice tip every 5–10 min. Within 20–45 min, the chromatin should turn into a fine suspension freely passing in and out of a regular 200 μL tip. When this is achieved, clear both fractions by centrifugation for 10 min at 20,000 × *g*, 4 °C, and transfer the supernatants from each fraction into a fresh tube. These supernatants contain the soluble nuclear proteins and the high molecular weight nuclear extract, respectively.

4 Notes

1. Cell numbers at 90% confluency vary by cell type. Mouse embryonic stem cells (mESC), which have a smaller cytoplasmic volume, can grow up to more than 20 million cells per 10 cm dish. HeLa cells, on the other hand, have relatively larger cytoplasm and yield less than five million cells on the same size of culture plate.
2. 0.025% trypsin has been diluted tenfold from what is typically used for passaging cells. Any growth culture medium containing FBS will work as an inhibitor of trypsin. If not possible to trypsinize the cells, scraping also works. In that case, scrape gently to avoid breaking cells. Damaged cells contaminate fractions in the following procedure. Generating a suspension of single cells is important in this step since detergent might not affect cells inside of a cell clump. If you notice cell clumps in the PBS/EDTA suspension, pass through a 0.4 μm cell strainer to remove aggregates and use single cells only.
3. Mix 10 μL of cell suspension with 740 μL of culture medium and 250 μL of 0.4% Trypan Blue. Make sure that viability is better than 90%.
4. Use low adhesion 1.5 or 1.7 mL tubes since certain cells tend to stick into the tube wall.
5. Higher centrifugation speed might spin down damaged cells that are not properly fractionated by the following procedure, and end up contaminating the fractions.
6. This step allows cells to swell by osmotic pressure. The salt concentration may need to be optimized depending on cell type. 15 mM KCl works on most cells. For neurons, whose nuclei break under these conditions, we recommend 50 mM NaCl. Do not resuspend the pellet by pipetting. Instead, flick very gently. Pellets that do not disperse completely are indicative of contamination with damaged cells.
7. The cytoplasmic membrane is dissolved at this step while the nuclear membrane remains intact. The final detergent concentration should be optimized on cells of your interest. For example, we use 0.0375% of Igepal CA-60 for mESC, 0.1% for neurons, and 0.125% for N2a cells as a final detergent concentration. For some durable cells such as HeLa the concentration should be increased to 0.15%. Prepare 5% Igepal CA-60 stock solution to make the desired 2× DCML buffer to minimize pipetting error due to high viscosity of the concentrated detergent. When optimizing conditions, use the same amount of cells and test different detergent concentrations. Incubation time with detergent should not be longer than 5 min from the

point of adding 2× DCML buffer to the beginning of centrifugation through the sucrose solution (**step 6**). The sample can be scaled up to 20 million cells per one 1.5 mL tube with an optimization of detergent concentration, while the other parameters are fixed. For example, the presence of cytoplasmic markers in the nucleoplasm and chromatin fractions is indicative of over-lysis, and requires lowering the detergent concentration.

8. Low adhesion tubes help to minimize the amount of nuclei stuck on the tube side wall while passing through the sucrose solution. The volume of this solution is 2.5 times the volume of the lysate, which is ~500 μL in this scale. Centrifugation conditions are important since nuclei can be lysed at high speed.
9. There is ~750 μL of supernatant cytoplasmic fraction, but most of the cytoplasmic material is in the top ~300 μL. To avoid nuclear contamination in the cytoplasmic fraction, collect only the top 300 μL. The nuclear pellet should be whitish. A yellowish hue may indicate that cells are lysed insufficiently. In this system, mitochondria and endoplasmic reticulum (ER) co-migrate with nuclei, so protein markers for these organelles are expected to be seen in the nucleoplasmic fraction.
10. This is an optional step for minimizing the contamination of cytoplasmic material in the nuclei. Some cytoplasmic components that may stick to the tube side wall can be released and pelleted along with the nuclei in the following steps. Gently pipette the nuclei with a small volume of PBS/EDTA using a large-orifice 200 μL tip.
11. Pure nuclei should easily disperse completely in PBS/EDTA. If part of the pellet does not go into suspension, there is a possibility that over-lysis occurred during the previous step, and nuclei were partially damaged.
12. Do not quick spin.
13. After quick spin in glycerol buffer, nuclei come off the bottom and go on top of the buffer.
14. Clumps of nuclei that do not disperse in glycerol buffer in **step 9** may be due to partial lysis. To break these aggregates, treat with 5 units of DNase I for 3–5 min at RT. For this, just add DNase I directly to nuclear suspension in glycerol buffer. Flick gently a few times during the incubation. When the aggregates are broken, place the tube briefly on ice to stop the reaction and proceed to the next step immediately.
15. Incubation for 2 min in the presence of Igepal CA-60 and 1 M urea strips loosely associated soluble material from chromatin [13]. The longer the incubation, the stronger the stripping.

When performing this step, always stick to one and the same incubation time for consistent results.

16. Two aliquots of chromatin pellet will be prepared in parallel by transferring 50% of nuclear lysate in this step. RNA or proteins can be prepared separately from these aliquots. The chromatin pellet is semi-transparent. Nucleoli will coprecipitate with chromatin, so fibrillarin can be detected in the chromatin fraction.
17. This washing step aims to remove the residual nucleoplasmic contents. The chromatin pellet will not be dissolved.
18. It can be a bit difficult to extract proteins and RNA from the chromatin pellet since it is very intact. For protein extraction, add 200 μL of 5% SDS lysis buffer directly to the pellet in one of the tubes and incubate for at least 15 min at 55 °C with vigorous vortexing every 5 min. For RNA extraction, add 200 μL of Trizol directly to the pellet in the other tube and vortex vigorously until pellet is completely dissolved. Incubation in a heat block at 50 °C will shorten the time.
19. Use the same percentage of each fraction when checking for purity, i.e., 5%. This corresponds to 15 μL of the cytoplasm (total volume of 300 μL), 10 μL of nucleoplasm (200 μL), and 20 μL of the chromatin fraction (two aliquots, 200 μL each). Test with antibodies specifically recognizing protein markers for each fraction.
20. It is possible to detect Histone proteins in cytoplasmic fraction of highly proliferative cell lines (e.g., mESC).
21. Extract RNA immediately to minimize degradation. Follow manufacturers' instructions for Trizol and Trizol LS use. Fractionation purity can also be checked by gene-specific RT-PCR. For this, treat with DNase I for 30 min prior to reverse transcription with GAPDH-specific antisense primer followed by PCR with GAPDH primers set to amplify a short intron and its flanking exons. The unspliced transcript should be enriched in chromatin fraction while the spliced product should be in the cytoplasmic fraction.
22. The buffer conditions are optimized for mammalian brain samples as well as for mouse neuroblastoma N2a and human embryonic kidney HEK293 cell lines. Other tissues or cell lines may need adjusting the sucrose concentration in the homogenization and cushion buffers.
23. Using more than nine volumes of homogenization buffer will produce homogenate with higher density than the cushion buffer. If the buffer to sample ratio has to be increased, the sucrose should be lowered proportionally.

24. The total volume of the homogenate (homogenization buffer plus sample) should not exceed 80% of the capacity of the grinder. The friction force during this process is considerable. For this reason, the use of grinders with a capacity larger than 30 mL is not recommended.
25. If the tube is less than 3/4 full, the wall above the homogenate may collapse during centrifugation. To prevent this, increase the volume by overlaying the homogenate with wash buffer, or cut the empty tube approximately 1 cm above the expected level of the homogenate.
26. The nuclei should lyse easily during this process, a fine chromatin precipitate will form and will start growing. However, some cell lines have fragile nuclei that disintegrate during homogenization in high sucrose. If this occurred, the pellet will contain aggregated chromatin that will stay in a clump impossible to resuspend.
27. The Benzonase requires Mg^{2+} ions. Extraction with DNase I instead of Benzonase is very inefficient since the DNase is inhibited by salt and has 50–100 fold lower activity in the lysis buffer than in the DNase I buffer.
28. Extraction with RNAases works efficiently only with high molecular weight pellets of volumes up to 5 μL. Larger-size pellets cannot be resuspended using RNase.

Acknowledgment

We thank Julia Nikolic for critical reading of the manuscript and Douglas L. Black for his support. K-H.Y. was supported by a fellowship from UCLA Broad Stem Cell Research Center's Training Program.

References

1. Xu L, Massagué J (2004) Nucleocytoplasmic shuttling of signal transducers. Nat Rev Mol Cell Biol 5:209–219
2. Blee TKP, Gray NK, Brook M (2015) Modulation of the cytoplasmic functions of mammalian post-transcriptional regulatory proteins by methylation and acetylation: a key layer of regulation waiting to be uncovered? Biochem Soc Trans 43:1285–1295
3. Jeffries CD, Fried HM, Perkins DO (2011) Nuclear and cytoplasmic localization of neural stem cell microRNAs. RNA 17:675–686
4. Liang H, Zhang J, Zen K, Zhang C-Y, Chen X (2013) Nuclear microRNAs and their unconventional role in regulating non-coding RNAs. Protein Cell 4:325–330
5. Miescher F (1871) In Hoppe-Seyler's Medizinischchemiche Untersuchungen. Hirschwald A (author) Berlin 4:441–460
6. Pandya-Jones A, Black DL (2009) Co-transcriptional splicing of constitutive and alternative exons. RNA 15:1896–1908
7. Bhatt DM, Pandya-Jones A, Tong A-J, Barozzi I, Lissner MM, Natoli G, Black DL, Smale ST (2012) Transcript dynamics of proinflammatory genes revealed by sequence analysis of subcellular RNA fractions. Cell 150:279–290

8. Damianov A, Ying Y, Lin C-H, Lee J-A, Tran D, Vashisht AA, Bahrami-Samani E, Xing Y, Martin KC, Wohlschlegel JA, Black DL (2016) Rbfox proteins regulate splicing as part of a large Multiprotein complex LASR. Cell 165:606–619
9. Grabowski PJ (2005) Splicing-active nuclear extracts from rat brain. Methods 37: 323–330
10. Sharma S, Falick AM, Black DL (2005) Polypyrimidine tract binding protein blocks the 5′ splice site-dependent assembly of U2AF and the prespliceosomal E complex. Mol Cell 19:485–496
11. Markovtsov V, Nikolic JM, Goldman JA, Turck CW, Chou MY, Black DL (2000) Cooperative assembly of an hnRNP complex induced by a tissue-specific homolog of polypyrimidine tract binding protein. Mol Cell Biol 20:7463–7479
12. Chou MY, Rooke N, Turck CW, Black DL (1999) hnRNP H is a component of a splicing enhancer complex that activates a c-src alternative exon in neuronal cells. Mol Cell Biol 19:69–77
13. Wuarin J, Schibler U (1994) Physical isolation of nascent RNA chains transcribed by RNA polymerase II: evidence for cotranscriptional splicing. Mol Cell Biol 14:7219–7225

Chapter 13

Isolation of Newly Transcribed RNA Using the Metabolic Label 4-Thiouridine

Angela Garibaldi, Francisco Carranza, and Klemens J. Hertel

Abstract

Isolation of newly transcribed RNA is an invaluable approach that can be used to study the dynamic life of RNA *in cellulo*. Traditional methods of whole-cell RNA extraction limit subsequent gene expression analyses to the steady-state levels of RNA abundance, which often masks changes in RNA synthesis and processing. This chapter describes a methodology with low cytotoxicity that permits the labeling and isolation of nascent pre-mRNA in cell culture. The resulting isolate is suitable for use in a series of downstream applications aimed at studying changes in RNA synthesis, processing, or stability.

Key words 4sU, 4sU-seq, Mammalian cells, Nascent RNA, Decay, Transcription, Nascent pre-mRNA, mRNA processing, Metabolic labeling, 4-Thiouridine

1 Introduction

The majority of gene expression research focuses on RNA transcript abundance at a steady-state level, providing only a snapshot of the cellular state. This glimpse of transcript abundance in the cell limits the understanding of regulation to whether a gene is generally up or down regulated. This obscures whether a change in gene expression is due to differences in the rate of transcription, the rate of degradation, or both. Previous approaches aimed at elucidating the dynamics of cotranscriptional pre-mRNA processing focused on a variety of immunoprecipitation and cell fractionation techniques following a chosen pathway induction (LPS stimulation) [1–3]. Likewise, pulse-chase experiments using well-known transcription inhibitors such as Actinomycin D have been frequently used to measure mRNA stability and degradation [2, 4]. While providing critical advances to the fundamental understanding

Angela Garibaldi and Francisco Carranza contributed equally to this work.

Yongsheng Shi (ed.), *mRNA Processing: Methods and Protocols*, Methods in Molecular Biology, vol. 1648,
DOI 10.1007/978-1-4939-7204-3_13, © Springer Science+Business Media LLC 2017

Table 1
Recommended 4sU concentrations [5]

Duration of labeling [min]	Recommended 4sU concentration [μM]
120	100–200
60	200–500
15–30	500–1000
<10	500–20,000

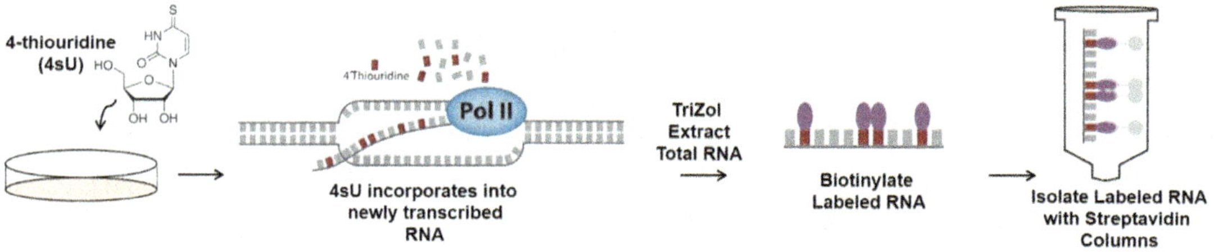

Fig. 1 Conceptual workflow of 4sU labeling and isolation of newly transcribed RNA

of RNA dynamics, these methods are limited by cytotoxicity and a lack of kinetic resolution [4]. The emergence of next-generation sequencing technologies, in conjunction with the uridine analog 4-thiouridine (4sU) as a metabolic label, has opened up an exciting avenue to studying genome-wide RNA kinetics at high resolution [5–7].

4sU can be used to metabolically label and track an RNA from synthesis to degradation by simply adding 4sU to mammalian cell culture media. 4sU is immediately taken up by cells, phosphorylated, and incorporated into any newly transcribed RNA. 4sU-labeled RNA can be tracked and isolated to study nascent RNA behavior using RNA sequencing. Alternatively, media replacement after a longer incubation period in 4sU media can be used to study the half-life and degradation of an RNA. Depending on the experimental approach taken, appropriate concentrations of 4sU should be selected for various cell types and incubation times to minimize off-target effects (*see* Table 1) [8].

Once whole-cell RNA is extracted, the 4sU-labeled RNA can be biotinylated via its sulfylhydryl group and selectively isolated using streptavidin-coated magnetic beads. Given the strong biotin/streptavidin interaction, 4sU labeled RNA can be stringently washed. Eluted 4sU labeled RNA can then be used in subsequent qRT-PCR and RNA-seq experiments, with or without ribosomal RNA depletion (*see* Fig. 1). Given the fact that 4sU can be used to mark nascent transcripts, the use of the 4sU labeling protocol can lead to a wealth of new findings that directly relate to immediate changes in gene expression.

2 Materials

All materials must be sterile, RNase-free, molecular biology grade. Large quantities of TRIzol reagent may be used for experiments. In case of contact with skin/eyes, have a polyethylene glycol 300 or 400 in industrial methylated spirits (70:30) solution prepared before proceeding.

2.1 4sU Labeling of Cells

1. 4-Thiouridine dissolved in sterile RNase-free water to 50 mM. Store in small aliquots at −20 °C, thawing only once.

2.2 Total RNA Extraction

1. TRIzol.
2. 75% EtOH (cthanol).
3. RNase-free water.
4. RNA Precipitation Solution: 0.8 M NaCl, 1.2 M NaCitrate.
5. (*Optional*) TE: 10 mM Tris, 1 mM EDTA.

2.3 Biotinylation of 4sU-Labeled RNA

1. EZ-Link Biotin-HPDP. Make stock aliquots 1 mg/mL dissolved in Dimethylformamide (*see* **Note 1**) and store at 4 °C.
2. 10× Biotinylation Buffer: 100 mM Tris pH 7.4, 10 mM EDTA. Store in aliquots of ~1 mL at 4 °C.
3. 5 M NaCl.
4. 75% EtOH.
5. (*Optional*) Phase Lock Gel Heavy Tubes (2.0 mL).

2.4 Separation of Labeled and Unlabeled RNA Using Streptavidin-Coated Magnetic Beads

1. μMacs Streptavidin Kit (*see* **Note 2**).
2. 1× Washing Buffer: 100 mM Tris pH 7.5, 10 mM EDTA, 1 M NaCl, 0.1% Tween20.
3. 100 mM Dithriothreitol (DTT) in RNase-free water.
4. Magnetic Separator and Stand (2 each). Alternatively, one of each is included in the starter kit.
5. (*Optional*) RNeasy MinElute Cleanup Kit.

2.5 Recovery of Unlabeled, Unbound RNA

1. Phenol/chloroform pH 6.7.
2. Isopropanol.
3. EtOH.

3 Methods

3.1 4sU Labeling of Nascent RNA

1. Plate a number of cells of the desired cell type in either a 10 cm or 15 cm tissue culture plate that will reach 70–80% confluency after 24 h. For a 10 cm plate, use at least 10 mL of culture medium. For a 15 cm plate, use at least 20 mL of culture medium.

2. For a 10 cm dish, a minimum of 5 mL of culture medium containing 4sU is needed. For a 15 cm dish, a minimum of 10 mL of culture medium containing 4sU is needed.
3. Once cells reach 70–80% confluency, transfer 5 mL or 10 mL of culture medium from the plate to a clean 15 mL conical tube.
4. Add 4sU to the culture medium in the conical tube and pipette up and down with a serological pipette to mix thoroughly. Refer to Table 1 for general guidelines for 4sU concentrations (*see* **Note 3**).
5. Aspirate the remaining unlabeled culture medium from the plate. Add the culture medium containing 4sU to the cells (*see* **Note 4**).
6. Incubate cells with 4sU culture medium for the desired amount of time. A longer incubation period is recommended for RNA decay studies (*see* **Note 5**).
7. Quench the reaction by quickly aspirating the 4sU culture medium and adding 3 mL of TRIzol for 10 cm plate, or 5 mL of TRIzol for 15 cm dishes.
8. Ensure the entire plate is covered by TRIzol and allow it to sit for 2–5 min for complete cell lysis.
9. Pipette the cell/TRIzol lysate to homogenize the cells and get all cells off the plate. Transfer the lysate to a 15 mL conical tube.
10. Immediately extract total RNA from TRIzol samples, or store at −80 °C between 6 months and 1 year (*see* **Note 6**).

3.2 Total RNA Extraction

1. Transfer 1 mL of TRIzol sample to each of (3) 1.5 mL microcentrifuge tubes.
2. Add 0.2 mL chloroform per mL TRIzol and shake vigorously for 15 s.
3. Incubate at room temperature for 2–3 min.
4. Centrifuge at 20,000 × *g* for 15 min at 4 °C (*see* **Note 7**).
5. Transfer aqueous upper phase (containing the RNA) to a new tube.
6. Add ½ the reaction volume of both RNA precipitation buffer and isopropanol (e.g., to 3 mL of supernatant add 1.5 mL RNA Precipitation Solution and 1.5 mL isopropanol).
7. Invert to mix well.
8. Incubate at room temperature for 10 min.
9. Centrifuge at 20,000 × *g* for 10 min at 4 °C.
10. Immediately remove the supernatant.
11. Wash with an equal volume of 75% EtOH.

12. Centrifuge at 20,000 × *g* for 10 min at 4 °C.
13. Immediately remove the supernatant.
14. Centrifuge again briefly to spin down remaining EtOH.
15. Remove remaining ethanol by pipetting using 200 μL pipette. Repeat step using 20 μL pipette (*see* **Note 8**).
16. Add 100 μL of 1× TE or RNase-free water per 100 μg expected RNA yield.
17. If needed, dissolve RNA by heating to 65 °C for 10 min.
18. Use a NanoDrop spectrophotometer to measure RNA yield. This RNA can be stored at −80 °C for at least 3 months with minimal freeze-thaws.

3.3 Biotinylation of 4sU-Labeled RNA

1. Labeling Reaction (use 60–100 μg total RNA):

   2 μL Biotin-HPDP (1 mg/mL DMF) per 1 μg RNA.

   1 μL 10× Biotinylation Buffer per 1 μg RNA.

   Bring up to7 μL with RNase-free water per 1 μg RNA.
2. Rotate at room temperature in the dark for at least 1.5 h (*see* **Note 9**).
3. Add an equal volume of Phenol/Chloroform pH 6.7.
4. Mix vigorously by vortex or by manually shaking.
5. Incubate for 2–3 min at room temperature until phases begin to separate and bubbles start to disappear.
6. Centrifuge at full speed (20,000 × *g*) for 5 min at 4 °C.
7. Carefully transfer upper phase into new tubes (*see* **Note 10**).
8. *RNA precipitation*: Add 1/10 the reaction volume of 5 M NaCl.
9. Add an equal volume of isopropanol, invert to mix well.
10. Centrifuge at 20,000 × *g* for 20 min at 4 °C.
11. Remove the supernatant. Add an equal volume of 75% EtOH.
12. Centrifuge at 20,000 × *g* for 10 min at 4 °C.
13. Remove EtOH completely and resuspend the RNA pellet at approximately 1 μg/μL with RNase-free water or TE.

3.4 Separation of Labeled and Unlabeled RNA Using Streptavidin-Coated Magnetic Beads

1. Heat biotinylated RNA samples to 65 °C for 10 min and immediately place on ice for 5 min.
2. Add up to 100 μg (max. 100 μL) of biotinylated RNA to 100 μL of streptavidin beads (*see* **Note 11**).
3. Incubate at room temperature with rotation for 15 min.
4. Place μMacs columns into magnetic stand. Process no more than eight samples at a time (*see* **Note 12**).

5. Add 0.9 mL of washing buffer to columns to prerun and equilibrate (*see* **Note 13**).
6. Apply bead-bound RNA to the columns.
7. For recovery of unlabeled/unbound RNA, collect this flow-through and *see* Subheading 3.5. Otherwise, discard the flow-through.
8. Place tubes or alternative collection apparatus underneath columns to catch the wash flow-through.
9. Wash 3× with 0.9 mL 65 °C washing buffer. *Optional:* For recovery of unlabeled RNA, collect the first wash and *see* Subheading 3.5.
10. Wash 3× with 0.9 mL room temperature washing buffer.
11. Elute the labeled RNA by placing the 1.5 mL microcentrifuge tubes underneath the columns and adding 100 μL 100 mM DTT to the columns (*see* **Note 14**).
12. Perform a second DTT elution into the same tubes 3–5 min later.
13. Immediately perform EtOH precipitation with 2.5 V 100% EtOH and 10 μg glycogen.
14. Precipitate overnight at −20 °C.
15. Spin at 20,000 × *g* for 15 min at 4 °C.
16. Wash with 75% EtOH.
17. Spin at 20,000 × *g* for 5 min at 4 °C.
18. Remove all EtOH using technique used in Subheading 3.2, **step 15** and resuspend in ~30 μL RNase-free water.
19. Spec labeled RNA with NanoDrop (*see* **Note 15**).

3.5 Recovery of Unlabeled, Unbound RNA (Optional)

1. For recovery of >90% of unbound RNA, collect the flow-through and the first wash for subsequent precipitation.
2. Combine the two fractions and recover the unbound RNA by isopropanol/EtOH precipitation as performed after the biotinylation reaction (*see* Subheading 3.3). Omit the addition of NaCl; the washing buffer has sufficient NaCl.

3.6 Validation

1. Validate with RT-PCR/qPCR by comparing labeled RNA to total RNA or unlabeled RNA for genes/transcripts of interest.

4 Notes

1. Gentle warming will ensure complete solubilization. Store aliquots at 4 °C. Alternatively, store 20 mg/mL at −20 °C. Do not use any polystyrene serological pipette in this process as DMF will degrade the plastic, leading to plastic residues that may inhibit biotinylation.

2. Per conversations with Miltenyi tech support, the beads are subject to the expiration date on the box. Columns, however, are good for 3 years. At the time of publication, beads are not sold separately.
3. Thaw 4sU only once, and just before use. Concentrations should be optimized based on cell line and desired labeling time to balance incorporation efficiency and possible inhibition of rRNA synthesis [8].
4. Handle labeled cells at room temperature as quickly as possible. Note that 4sU has crosslinking ability at 365 nm wavelength. Avoid light sources that may mimic this wavelength.
5. To study RNA decay you can perform a pulse chase experiment in which the duration of the 4sU labeling is increased and chased with cell media absent of 4sU. Timepoints can then be taken during the chase period to determine decay rates.
6. TRIzol samples may be freeze-thawed at least twice, thus allowing for two pull-down reactions on different dates from a single 15 cm plate depending on cell type. Otherwise, freeze in two aliquots to reduce freeze-thaws.
7. While not "best practice," centrifugation at room temperature will not cause failure.
8. After these two steps, no further drying of the pellet is required. Over drying of pellet may risk making it difficult to dissolve, even with heating.
9. Rotation has been done under general lab lighting with success.
10. Alternatively, this step can be done using phase lock gel heavy tubes to avoid both the loss of material and phenol carry-over.
11. 80 μL of beads for 80 μg RNA reaction is also sufficient.
12. When processing replicate samples, we find increased variability when the pulldown is done in different rounds. Therefore, it is recommended to perform the pulldown on replicates in the same round.
13. To initiate the flow through the column you can gently press on the top of the column with your gloved finger.
14. Here you have the option to finish the remainder of this section by eluting directly into 700 μL RLT buffer and complete RNA isolation/cleanup using RNeasy MinElute cleanup kit. However, residual kit buffer in the RNA may skew NanoDrop OD readings.
15. For very short time points, this may be very low or unreliable detection.

Acknowledgments

Research in the Hertel laboratory is supported by NIH (GM062287, GM110244 and F31CA17179). Special thanks to Nate Hoverter for contribution of key graphics in Fig. 1.

References

1. Pandya-Jones A, Black DL (2009) Co-transcriptional splicing of constitutive and alternative exons. RNA 15(10):1896–1908
2. Core LJ, Waterfall JJ, Lis JT (2008) Nascent RNA sequencing reveals widespread pausing and divergent initiation at human promoters. Science 322(5909):1845–1848
3. Brody Y, Neufeld N, Bieberstein N, Causse SZ, Böhnlein E-M, Neugebauer KM et al (2011) The in vivo kinetics of RNA polymerase II elongation during co-transcriptional splicing. PLoS Biol 9(1):e1000573
4. Tani H, Akimitsu N (2012) Genome-wide technology for determining RNA stability in mammalian cells. Historical perspective and recent advantages based on modified nucleotide labeling. RNA Biol 9(10):1233–1238
5. Rädle B, Rutkowski AJ, Ruzsics Z, Friedel CC, Koszinowski UH, Dölken L (2013) Metabolic labeling of newly transcribed RNA for high resolution gene expression profiling of RNA synthesis, processing and decay in cell culture. J Vis Exp 78:e50195
6. Rabani M, Levin JZ, Fan L, Adiconis X, Raychowdhury R, Garber M et al (2011) Metabolic labeling of RNA uncovers principles of RNA production and degradation dynamics in mammalian cells. Nat Biotechnol 29(5):436–442
7. Barrass JD, Reid JEA, Huang Y, Hector RD, Sanguinetti G, Beggs JD et al (2015) Transcriptome-wide RNA processing kinetics revealed using extremely short 4tU labeling. Genome Biol 16:282
8. Burger K, Mühl B, Kellner M, Rohrmoser M, Gruber-Eber A, Windhager L et al (2013) 4-thiouridine inhibits rRNA synthesis and causes a nucleolar stress response. RNA Biol 10(10):1623–1630

Chapter 14

Robust, Cost-Effective Profiling of RNA Binding Protein Targets with Single-end Enhanced Crosslinking and Immunoprecipitation (seCLIP)

Eric L. Van Nostrand, Thai B. Nguyen, Chelsea Gelboin-Burkhart, Ruth Wang, Steven M. Blue, Gabriel A. Pratt, Ashley L. Louie, and Gene W. Yeo

Abstract

Profiling of RNA binding protein targets in vivo provides critical insights into the mechanistic roles they play in regulating RNA processing. The enhanced crosslinking and immunoprecipitation (eCLIP) methodology provides a framework for robust, reproducible identification of transcriptome-wide protein-RNA interactions, with dramatically improved efficiency over previous methods. Here we provide a step-by-step description of the eCLIP method, along with insights into optimal performance of critical steps in the protocol. In particular, we describe improvements to the adaptor strategy that enables single-end enhanced CLIP (seCLIP), which removes the requirement for paired-end sequencing of eCLIP libraries. Further, we describe the observation of contaminating RNA present in standard nitrocellulose membrane suppliers, and present options with significantly reduced contamination for sensitive applications. These notes further refine the eCLIP methodology, simplifying robust RNA binding protein studies for all users.

Key words RNA binding protein, CLIP-seq, eCLIP, seCLIP-seq, seCLIP, CLIP, eCLIP, RNA genomics

1 Introduction

RNA processing has been shown to play pivotal roles in shaping the cellular landscape through regulation of both protein-coding RNAs as well as regulation of processing and function of multiple classes of noncoding RNAs including long intergenic noncoding RNAs (lincRNAs) and small RNAs including microRNAs and piR-

The original version of this chapter was revised. The erratum to this chapter is available at: DOI 10.1007/978-1-4939-7204-3_19

Eric L. Van Nostrand and Thai B. Nguyen contributed equally to this work.

Yongsheng Shi (ed.), *mRNA Processing: Methods and Protocols*, Methods in Molecular Biology vol. 1648, DOI 10.1007/978-1-4939-7204-3_14, © The Author(s) 2017

NAs among others [1–3]. These regulatory steps have been shown to play critical roles across a variety of developmental stages, and misregulation of RNA processing has been implicated in many human diseases, including cancer [2, 3]. Orchestrating these RNA processing and regulatory roles are RNA binding proteins (RBPs), which play a variety of roles including controlling alternative splicing of mRNA transcripts, targeting RNAs to specific organelles or subcellular localizations with the cell, controlling RNA stability and turnover, and defining the timing and rate of translation [4]. Recent studies have indicated that there are over 1500 RBPs, and this number is continuing to expand with additional studies [5]. Despite these known important roles, however, detailed studies to describe the targets of and regulatory mechanisms have largely focused on a small set of RBPs, with the majority of RBPs remaining poorly characterized.

Recent advancements in next-generation sequencing have made it possible to study RBPs and their RNA targets in an unbiased and transcriptome-wide manner [6, 7]. Building upon early RNA ImmunoPrecipitation (RIP) approaches that identified protein binding to entire transcripts, CrossLinking and ImmunoPrecipitation (CLIP) enabled high-resolution profiling of binding sites [8]. In CLIP, RNA-protein interactions are stabilized via ultraviolet crosslinking, a desired protein is immunoprecipitated using a factor-specific antibody, and associated RNA is isolated and converted into DNA library suitable for high-throughput sequencing [8]. Various modifications of CLIP have since been described, including the use of photoactivatable nucleoside analogs (PNAs) to improve crosslinking efficiency (PAR-CLIP) [9] and computational and experimental methods to identify binding with single-nucleotide resolution. Proteinase K treatment of UV-crosslinked protein-RNA complexes leaves at least one amino acid covalently crosslinked to its associated ribonucleotide. Reverse transcriptase enzymes can create deletions at these positions [10] or, more often, terminate elongation due to the inability to read through this coupling, leading to a substantial fraction of cDNA fragments that terminate at the position of crosslinking. By incorporating a circular ligation step, iCLIP positions this crosslinking site at the start of the sequencing reads to enable identification of binding sites with single-nucleotide resolution [11]. However, widespread usage of these methods has been limited by the low efficiency of converting RNA molecules into adapter-ligated library, which leads to high experimental failure rates and high wasted sequencing (often >90% of reads) due to the presence of substantial PCR duplication [12].

We recently described enhanced CLIP (eCLIP), which incorporated high-efficiency enzymatic steps to achieve thousand-fold improved library efficiency [12]. The improved efficiency dramatically decreases experimental failure rates and PCR duplication, and enabled quantitative comparison with paired size-matched input to remove common CLIP artifacts. Here, we describe a detailed protocol for seCLIP, a simplified, single-end version of the eCLIP methodology, as well as assorted notes for critical handling steps. By ligating an adapter at the 3′ end of the cDNA fragment, eCLIP (similar to iCLIP) utilizes this termination to enrich for read pile-ups at these sites. Due to the adapter strategy used in our initial eCLIP procedure that positioned the I7 adaptor at this site, eCLIP initially required paired-end sequencing to obtain these crosslink sites at the 5′ end of the second read, as well as the unique molecular identifier (UMI, or random-mer) used for PCR duplicate identification. In this protocol, we describe an altered adapter strategy to enable single-end sequencing for eCLIP experiments.

Additionally, during eCLIP experiments on cell types with low total RNA quantity, we observed that a substantial fraction (in some cases more than half) of sequenced reads did not map to the human genome and instead mapped to a single bacterial contamination. We trace this source of contamination to manufacturer-supplied nitrocellulose membranes, and describe alternate sources for this material that alleviate this contamination and a method to assay membranes for eCLIP suitability. With a new 3′ linker ligation strategy that allows for single-end sequencing, and the alleviation of contamination found commonly in major suppliers, seCLIP brings modifications to eCLIP that will allow for more cost-efficient sequencing as well as paving the way for future low-input RNA as starting material for CLIP-seq experiments.

2 Materials

2.1 Crosslinking of Cultured Cells

1. 1× DPBS.
2. 254 nM UV crosslinker.
3. Cell scraper.
4. Liquid Nitrogen.

2.2 seCLIP

1. **Lysis buffer**: 50 mM Tris–HCl pH 7.4, 100 mM NaCl, 1% NP-40 (Igepal CA630), 0.1% SDS, 0.5% sodium deoxycholate (protect from light), 1:200 Protease Inhibitor Cocktail III (add fresh), in RNase/DNase-free H_2O.
2. Protease Inhibitor Cocktail III.
3. DNase.
4. RNase I.
5. RNase Inhibitor.
6. Dynabeads M-280 sheep anti-rabbit or Protein A/G magnetic beads.
7. **High salt wash buffer**: 50 mM Tris–HCl pH 7.4, 1 M NaCl, 1 mM EDTA, 1% NP-40, 0.1% SDS, 0.5% sodium deoxycholate (protect from light), in RNase/DNase-free H_2O.

8. **Wash buffer**: 20 mM Tris–HCl pH 7.4, 10 mM $MgCl_2$, 0.2% Tween-20, in RNase/DNase-free H_2O.
9. **1× TAP Buffer**: 10 mM Tris pH 7.5, 5 mM $MgCl_2$,100 mM KCl, 0.02% Triton X-100, in RNase/DNase-free H_2O.
10. Thermosensitive Alkaline Phosphatase (TAP) (1 unit/μL).
11. **5× PNK pH 6.5 buffer**: 350 mM Tris–HCl pH 6.5, 50 mM $MgCl_2$, in RNase/DNase-free H_2O.
12. 0.1 M DTT.
13. T4 PNK.
14. **1× RNA Ligase Buffer**: 50 mM Tris–HCl pH 7.5, 10 mM $MgCl_2$, in RNase/DNase-free H_2O.
15. 10× Ligase Buffer without DTT.
16. 0.1 M ATP.
17. 100% DMSO.
18. 50% PEG 8000.
19. T4 RNA ligase 1 high concentration.
20. 4–12% Bis-Tris Gel.
21. NuPAGE 4× LDS Sample Buffer.
22. NuPAGE MOPS SDS Running Buffer 20×.
23. NuPAGE Transfer Buffer 20×.
24. PVDF membrane.
25. Nitrocellulose membrane (*see* **Note 4**):

(a) iBlot 2 Transfer Stacks	ThermoFisher	IB23001	lot #2NR26016-01

or

(b) Amersham Protran Premium	GE	1060008	lot #G9931040

26. 5% milk + TBST (1× TBS pH 7.4 + 0.05% Tween-20).
27. Rabbit TrueBlot HRP secondary antibody.
28. ECL Western Blotting detection assay.
29. Proteinase K.
30. Urea.
31. Acid Phenol/Chloroform/Isoamylalcohol pH 4.5.
32. Phase lock heavy 2 mL Tubes.
33. 100% Ethanol.
34. RNA Clean & Concentrator-5 Kit.
35. Dynabeads MyOne Silane.
36. RLT Buffer.
37. 5 M NaCl.

38. 10× Ligase Buffer with DTT.
39. 10× AffinityScript reverse transcriptase buffer.
40. AffinityScript reverse transcriptase.
41. dNTPs (25 mM each).
42. Exo-SAP-IT.
43. 0.5 M EDTA.
44. 1 M NaOH.
45. 1 M HCl.
46. 5 mM Tris–HCl pH 7.5.
47. 10 mM Tris–HCl pH 7.5.
48. Q5 or other high fidelity PCR Master Mix.
49. qPCR Master Mix.
50. Agencourt AMPure XP beads.
51. MinElute gel purification Kit.
52. D1000 DNA tape/reagent.

2.3 Contamination Assay

1. TRIzol® Reagent.
2. TRIzol® LS Reagent.
3. SuperScript II (200 unit/μL).

2.4 Primer Sequences

seCLIP:

1. InvRiL19: /5Phos/rArGrArUrCrGrGrArArGrArGrCrArCrA rCrGrUrC/3SpC3/
   (Order 100 nmole RNA oligo, standard desalting; storage stock 200 μM; working stock 40 μM; final concentration 1 μM (input), 4 μM (CLIP)).
2. InvRand3Tr3: /5Phos/NNNNNNNNNNAGATCGGAAGA GCGTCGTGT/3SpC3/
   (Order 100 nmole DNA oligo, standard desalting; storage stock 200 μM; working stock 80 μM; final concentration 3 μM).
3. InvAR17: CAGACGTGTGCTCTTCCGA (25 nmole DNA oligo, standard desalting; storage stock 200 μM; working stock 20 μM; final concentration 0.5 μM).
4. D5x_qPCR: AATGATACGGCGACCACCGAGATCTACAC TATAGCCTACACTCTTTCCCTACACGACGC TCTTCCGATCT.
5. D7x_qPCR: CAAGCAGAAGACGGCATACGAGATCGAGT AATGTGACTGGAGTTCAGACGTGTGCTCTTCCGATC.

XBB1 contamination primers:

1. XBB1_qPCR_F: GAGGCGGCAAATATCCTGTG.
2. XBB1_qPCR_R: GTTTCACTTCCCCTCGTTCG.

3 Methods

See Fig. 1 for schematic of eCLIP methodology.

3.1 Crosslinking of Cultured Cells

3.1.1 Preparation of Cultured Cells for Crosslinking

- Aspirate the spent media and wash the plate with 1× DPBS (~15 mL for a 15 cm plate) at room temperature. For suspension cells, pellet cells to remove spent media, and wash once with 1× DPBS.
- Aspirate the supernatant from previous, and add enough 1× DPBS (~5 mL for a 15 cm plate) to cover the plate. For suspension cells, resuspend in 1× DPBS to a cell density of 2 × 10^7 cells per mL, place 3 mL of cell suspension in 1× DPBS on a 10 cm plate.

3.1.2 Crosslinking of Cultured Cells

- Place the tissue culture plate on ice or cooling block, and place the entire apparatus into the UV crosslinker, making sure the plate is leveled. Remove lid prior to crosslinking.
- Crosslink the cells with an energy setting of 400 mJ/Cm^2.
- While keeping the cells on ice, use a cell scraper to scrape the plate. Transfer the cells to a 50 mL conical tube. Wash the plate with 10 mL of 1× DPBS and add the wash to the 50 mL conical tube. Gently resuspend the cell until a homogenous mixture is obtained.
- Spin cells down at 200 × *g* for 5 min at room temperature.
- Aspirate the supernatant. Resuspend in the desired amount for flash freezing, typically 2 × 10^7 cells per mL.
- Transfer desired amount into 1.5 mL epi-tubes, and then spin down at 200 × *g* for 5 min at room temperature. Aspirate the supernatant and freeze by submerging the epi-tubes completely in liquid nitrogen. Store at −80 °C.

3.2 Cell Lysis and RNA Fragmentation

3.2.1 Lyse Cells

- Add 1 mL cold **lysis buffer** + 5.5 μL 200× **Protease Inhibitor Cocktail III** to each pellet (*see* **Note 1**).
- Pipette to resuspend, incubate for 15 min on ice.
  - At this time, begin antibody coupling (Subheading 3.3.1).

Fig. 1 Schematic of seCLIP method. (**a**) Crosslinking of cultured cells (Subheadings 3.1.1 and 3.1.2). (**b**) Lysis of crosslinked cells (Subheading 3.2.1). (**c**) RNA fragmentation with RNase (Subheading 3.2.2). (**d**) Immunoprecipitation of RBP-RNA complexes (Subheadings 3.3.1–3.3.5). (**e**) Dephosphorylation of RNA fragments and ligation of 3′ RNA adapter (Subheading 3.5.1). (**f**) Polyacrylamide gel electrophoresis and membrane transfer (Subheadings 3.6.1–3.6.5). (**g**) Mince preparative membrane into ~2 mm squares (Subheadings 3.6.6) (**h**) RNA isolation from membrane (Subheadings 3.6.7 and 3.6.8). (**i**) Dephosphorylation of RNA fragments and ligation of 3′ RNA adapter for input samples (Subheadings 3.7.1–3.8.3). (**j**) Reverse transcription of RNA (Subheadings 3.9.1 and 3.9.2). (**k**) cDNA cleanup (removal of excess primers and RNA) (Subheading 3.9.3). (**l**) Ligation of 3′ DNA adapter (on-bead) and cleanup (Subheadings 3.10.1–3.10.3). (**m**) PCR amplification of cDNA library and cleanup (Subheadings 3.11.2–3.12.2). (**n**) Final Structure of eCLIP library fragment. The unique molecular identifier or random-mer is shown in green and abbreviated as UMI

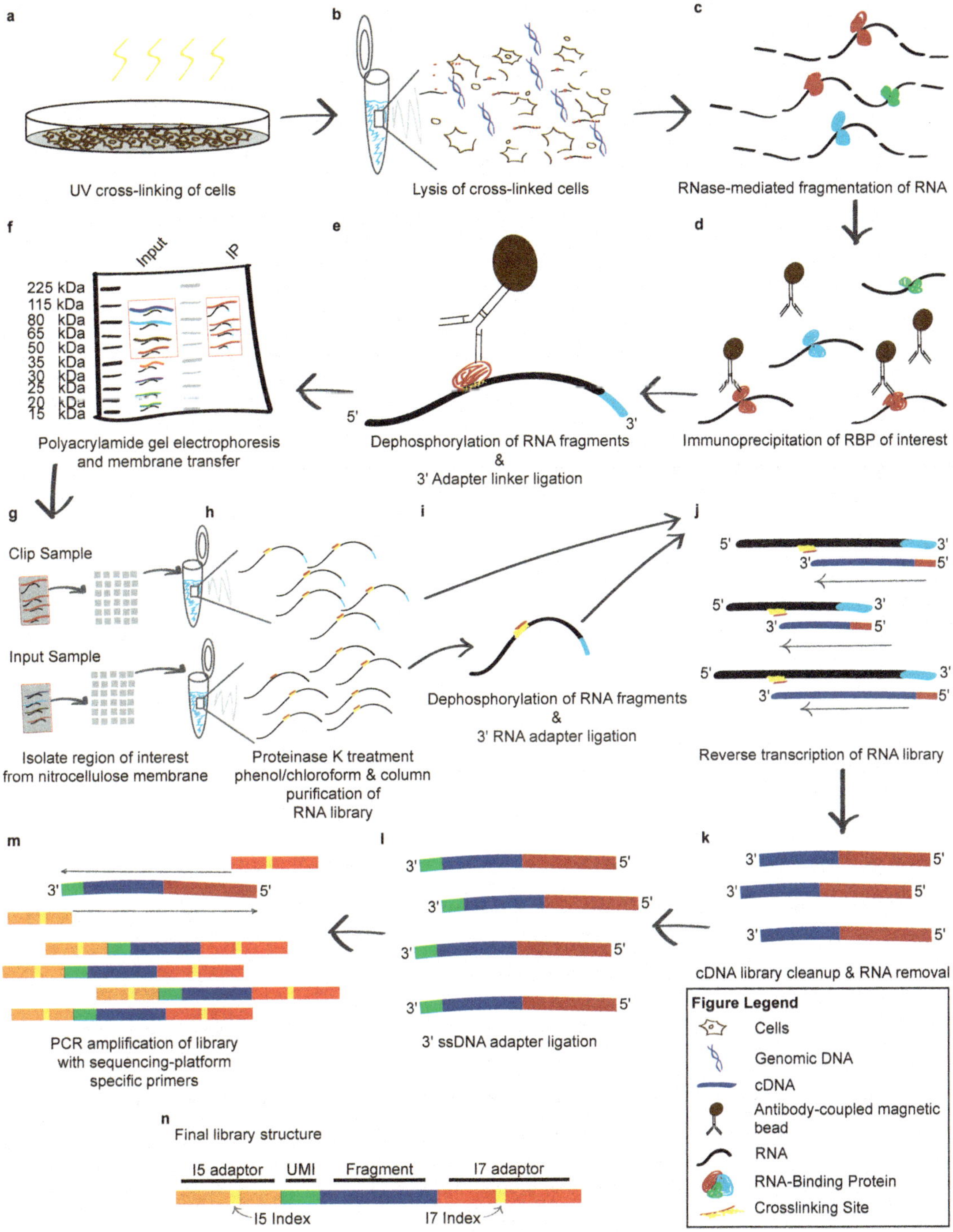

Fig. 1 (continued)

3.2.2 RNase Treat Lysate

- Sonicate in Bioruptor at "low" setting, 30 s on / 30 s off for 5 min at 4 °C.
- Add 2 μL **DNase** to lysate.
- Dilute RNase I 1:25 in 1× PBS (prechilled to 4 °C).
- Add 10 μL **diluted RNase I** to lysate, mix and immediately proceed to the next step.
- Incubate in Thermomixer at 1200 rpm, 37 °C for 5 min.
- Place on ice, add 11 μL **Murine RNase Inhibitor** and pipette mix (*see* **Note 1**).
- Centrifuge 15,000 × *g*, 4 °C for 15 min → Transfer the supernatant to a new tube and discard pellet.

3.3 Capture RBP-RNA Complexes on Beads

3.3.1 Couple Antibody to Magnetic Beads (Start During Subheading 3.2.1)

- Use **125 μL beads + 10 μg antibody** per sample.

3.3.2 Prewash Beads

- Magnetically separate beads, remove the supernatant.
- Wash beads 2× in 500 μL cold **lysis buffer.**
- Resuspend beads in 100 μL cold **lysis buffer.**

3.3.3 Bind Antibody

- Add 10 μg of antibody to 100 μL washed beads.
- Rotate at room temperature for 45 min.

3.3.4 Capture RBP-RNA Complexes on Beads

- Wash antibody beads 2× in 500 μL cold **lysis buffer.**
- Add the supernatant from Subheading 3.2.2 to washed antibody beads.
- Rotate at 4 °C overnight

3.3.5 Remove Input Samples

- To a new tube, take 20 μL (2%) of total beads + lysate sample for Preparative gel, store at 4 °C.
- To a new tube, take 20 μL (2%) of total beads + lysate sample for Imaging gel; store at 4 °C.

3.4 Immunoprecipitation Washes and RNA Dephosphorylation

Wash Beads (Pre-chill All Wash Buffers to 4 °C) (See Note 2)

- Magnetically separate beads, remove the supernatant (or store for western if desired).
- Wash 2× with 900 μL **High salt wash buffer**, then 1× with 500 μL **Wash buffer.** Remove the supernatant.
- Add 500 μL **Wash buffer**, mix, separate on a magnet, add 500 μL **1× TAP** buffer, mix, remove the supernatant. Wash 1× with 500 μL **1× TAP buffer,** remove the supernatant.

3.4.1 TAP Treat (on-Bead)

- **Prepare TAP master mix** on ice (100 μL per sample):

H_2O	79 μL
10× TAP buffer	10 μL
Murine RNase Inhibitor	2 μL
DNase	1 μL
TAP enzyme	8 μL

- Add **100 μL TAP mix** to each sample, incubate in Thermomixer at 1200 rpm, 37 °C for 15 min.

3.4.2 PNK Treat (On-Bead)

- **Prepare PNK master mix** on ice (300 μL per sample):

H_2O	224 μL
5× PNK pH 6.5 buffer	60 μL
0.1 M DTT	3 μL
Murine RNase Inhibitor	5 μL
DNase	1 μL
T4 PNK enzyme	7 μL

- Add **300 μL PNK mix** to each sample, incubate in Thermomixer at 1200 rpm, 37 °C for 20 min.

3.4.3 Wash Beads (Prechill Buffers to 4 °C)

- Magnetically separate beads, remove the supernatant.
- Wash with 1× 500 μL **Wash buffer,** remove the supernatant.
- Add 500 μL **Wash buffer**, mix, add 500 μL **High salt wash buffer**, mix, remove the supernatant.
- Add 500 μL **High salt wash buffer**, mix, add 500 μL **Wash buffer**, mix, remove the supernatant.
- Wash 1× with 500 μL **Wash buffer**, remove the supernatant.
- Add 500 μL **Wash buffer**, mix, add 300 μL **1× RNA Ligase buffer (no DTT)**, mix, remove the supernatant.
- Repeat wash 2× with 300 μL **1× RNA Ligase buffer (no DTT)**, carefully remove all remaining supernatant.

3.5 Ligate 3′ RNA Adapter (On-Bead)

3.5.1 Prepare RNA Adapter Ligation Master Mix

- Prepare on ice; 25 μL per sample:

H_2O	9 μL
10× Ligase buffer (*no* DTT)	3 μL
0.1 M ATP	0.3 μL

(continued)

100% DMSO	0.8 μL
50% PEG 8000	9 μL
Murine RNase Inhibitor	0.4 μL
High concentration T4 RNA Ligase	2.5 μL

- Mix carefully by pipetting (do not vortex) (*see* **Note 3**).

3.5.2 Perform RNA Adapter Ligation

- Add **25 μL RNA adapter ligation master mix** to each sample.
- Add **2.5 μL InvRiL19 RNA adapter** to each sample.
- Incubate at room temperature for 75 min; flick to mix every ~10 min.

3.6 Western Blotting and RNA Isolation from Membrane

3.6.1 Wash Beads (Prechill Buffers to 4 °C)

- Add 500 μL **Wash buffer**, magnetically separate, remove the supernatant.
- Add 500 μL **Wash buffer**, mix, add 500 μL **High salt wash buffer**, mix, remove the supernatant.
- Wash 1× with 500 μL **High salt wash buffer**, remove the supernatant.
- Add 500 μL **High salt wash buffer**, mix, add 500 μL **Wash buffer**, mix, remove the supernatant.
- Repeat wash 2× with 500 μL **Wash buffer.**

3.6.2 Prepare Samples for Gel Loading

- **Preparative gel**—used for membrane transfer and RNA isolation.
- **Imaging gel**—for IP-western blot imaging of IP success.
- **Remove supernatant,** add **100** μL cold **Wash buffer**, resuspend beads well.
- Move 20 μL (20% of sample) to new tube #1 for **Imaging gel IP samples**.
- Magnetically separate, remove the supernatant.
- Resuspend sample in 20 μL cold **Wash buffer** = **Preparative gel IP samples**.
- Thaw on ice **Preparative and Imaging input samples** saved at Subheading 3.3.5.
- Prepare each sample mix:

IP or input sample	20 μL
4× NuPAGE buffer	7.5 μL
1 M DTT	3.0 μL

- Denature all samples in Thermomixer (1200 rpm, 70 °C for 10 min), and then cool on ice >1 min.
- For **all samples**, magnetically separate and only load supernatant (IP AND Inputs have beads).

3.6.3 Load and Run Gels

- Load Preparative gel (4–12% Bis-Tris, 10-well, 1.5 mm) with (M) prestained markers and (m) diluted prestained marker (2 μL marker +2 μL 4× NuPAGE buffer +6 μL Wash Buffer) as follows for two experiments (A and B), leaving marker-only lanes between samples:

1	2	3	4	5	6	7	8	9	10
M	A-Input	(m)	A-IP	(m)	B-Input	(m)	B-IP	M	(m)

  - Load: 30 μL (all) volume for all preparative samples (input and IP).
- Load Imaging gel (4–12% Bis-Tris, 10 or 12-well, 1.5 mm).
  - Load 15 μL for all cold samples, save remaining volume at −20 °C as backup.
- Run at 150 V in 1× MOPS running buffer for 75 min or until dye front is at the bottom.

3.6.4 Transfer to Membranes

- **Prepare transfer:** (use 4 °C **transfer buffer**)
  - **Imaging gel**: Use PVDF membrane, activate with Methanol, and then move to **transfer buffer**.
  - **Preparative gel**: Use Nitrocellulose membrane (*see* **Note 4**), presoak in **transfer buffer**.
  - Assemble transfer stacks, from bottom to top (black side (negative) of stack holder on bottom):

    (negative)1× sponge – 2× Whatman – gel – membrane – 2× Whatman – 1× sponge (positive).
- **Transfer:**
  - Overnight at 30 V (preferred) or 2 h at 200 mA.
  - After removing **Preparative** membrane, rinse quickly once with sterile 1× PBS, then wrap in Saran wrap and store at −20 °C until Subheading 3.6.6.

3.6.5 Develop Imaging Membrane

- Block in 5% milk in TBST at room temperature for 30 min.
- Probe with primary antibody at appropriate concentration (typically 0.2 μg/mL) in 5% milk in TBST, at room temperature for 1 h.
- Wash 3× with TBST, 5 min.

- Probe with secondary antibody: 1:4000 Rabbit TrueBlot HRP in 5% milk in TBST, incubate at room temperature for 1–3 h.
- Wash 3× with TBST, 5 min.
- Develop with ECL, 30 s to 5 min, image with standard western blot film.

3.6.6 Cut Preparative Membrane (Cut Bands Based on Western Image)

- Place preparative membrane on clean glass/metal surface.
- Using a fresh razor blade, cut desired lane from the RBP band to 75 kDa above.
- Slice membrane piece into ~1–2 mm slices.
- Transfer slices to Eppendorf tube—place tube on ice if doing many samples.

3.6.7 Release RNA from Membrane

ProK mix on ice, 200 μL per sample: • PK buffer: 160 μL • Proteinase K: 40 μL	**Urea/PK buffer** • Dissolve 420 mg Urea in 500 μL PK buffer, then add PK buffer to a final volume of 1 mL

- Add **200 μL ProK mix** to membrane slices, incubate in Thermomixer (1200 rpm, 37 °C for 20 min).
- Add **200 μL Urea/PK buffer** to samples, mix, incubate in Thermomixer (1200 rpm, 37 °C for 20 min).

3.6.8 Purify RNA

- Add 400 μL **acid phenol/chloroform/isoamyl alcohol** (pH 4.5), mix well, incubate in Thermomixer at 1200 rpm, 37 °C for 5 min.
- Transfer all except membrane slices to Phaselock gel (Heavy) tube, incubate in Thermomixer at 1200 rpm, 37 °C for 5 min.
- Centrifuge at 13,000 × *g* for 15 min at room temperature.
- Transfer the aqueous (top) layer to a new 15 mL (or at least 3 mL volume) conical tube.
- Add 2 volumes RNA binding buffer (typically 2× ~400 = 800 μL).
- Add equal volume 100% ethanol and mix (typically ~1200 μL).
- Transfer 750 μL of mixed sample to Zymo-Spin column.
- Centrifuge for 30 s and discard flow-through.
- Repeat spins by reloading additional 750 μL volume until all sample has been spun through column.
- Add 400 μL RNA Prep Buffer, centrifuge for 30 s, discard flow-through.
- Add 700 μL RNA Wash Buffer, centrifuge for 30 s, discard flow-through.
- Add 400 μL RNA Wash Buffer, centrifuge for 30 s, discard flow-through.

- Centrifuge for additional 2 min.
- Transfer column to a new 1.5 mL tube (avoid getting wash buffer on column).
- Add 10 μL H_2O to column, let it sit for 1 min, centrifuge for 30 s.
- **Store CLIP samples at −80 °C until Subheading 3.9 (avoid multiple freeze-thaw cycles).**

3.7 Dephosphorylation of Input RNA

3.7.1 TAP Treat Input RNA

- **Prepare TAP master mix**

– H_2O	10 μL
– 10× TAP buffer	2.5 μL
– RNase Inhibitor	0.5 μL
– TAP enzyme	2.5 μL

- To **INPUT samples ONLY**, add 15.5 μL **TAP master mix.**
- Mix and incubate in Thermomixer at 1200 rpm, 37 °C for 15 min.

3.7.2 PNK Treat Input RNA

- Prepare PNK master mix (75 μL per sample).

– H_2O	45 μL
– 5× PNK pH 6.5 buffer	20 μL
– 0.1 M DTT	1 μL
– DNase	1 μL
– Murine RNase Inhibitor	1 μL
– T4 PNK enzyme	7 μL

- Add **75 μL** to samples and mix. Incubate in Thermomixer at 1200 rpm, 37 °C for 20 min.

3.7.3 Silane Cleanup Input RNA

- **Prepare beads:**
  - Magnetically separate 20 μL **MyONE Silane beads** per sample, remove the supernatant.
  - Wash 1× with 900 μL **RLT buffer.**
  - Resuspend beads in 300 μL **RLT buffer** per sample.
- **Bind RNA:**
  - Add beads in 300 μL **RLT buffer** to sample, mix.
  - Add 10 μL **5 M NaCl.**
  - Add 615 μL **100% EtOH.**
  - Mix, rotate at room temperature, 15 min.

- **Wash beads:**
  - Magnetically separate, remove the supernatant.
  - Add 1 mL **75% EtOH**, pipette resuspend and move suspension to **new tube.**
  - After 30 s, magnetically separate, remove the supernatant, and wash 2× with **75% EtOH** (30 s).
  - Magnetically separate, remove residual liquid with fine tip → air-dry 5 min.
- **Elute RNA:**
  - Resuspend in **10 μL H_2O**, let it sit for 5 min.
  - Magnetically separate and transfer 5 μL of supernatant to a new tube (for 3′ RNA adapter ligation below). Transfer the remainder of the supernatant to a new tube and store at −20 °C (as a backup input RNA sample).

3.8 3′ RNA Adapter Ligation to Input RNA

3.8.1 Anneal Adapter

- Take 5 μL of RNA (from above).
- Add 1.5 μL 100% DMSO and 0.5 μL **InvRiL19** adapter (*see* **Note 3**).
- Incubate at 65 °C for 2 min. Place on ice >1 min.

3.8.2 Prepare Ligation Master Mix; 13.5 μL Per Sample

10× Ligase Buffer (*with* DTT)	2.0 μL
0.1 M ATP	0.2 μL
Murine RNase Inhibitor	0.2 μL
100% DMSO	0.3 μL
50% PEG 8000	8.0 μL
RNA Ligase high conc	1.3 μL
H_2O	1.5 μL

- Add **13.5 μL** to each sample, mix, incubate at room temperature for 75 min. Flick to mix every ~15 min.

3.8.3 Silane Cleanup Input RNA

- **Prepare beads:**
  - Magnetically separate 20 μL **MyONE Silane beads** per sample, remove the supernatant.
  - Wash 1× with 900 μL **RLT buffer** and resuspend beads in 61.6 μL **RLT buffer.**
- **Bind RNA:**
  - Add beads in 61.6 μL **RLT buffer** to sample, mix and add 61.6 μL **100% EtOH.**
  - Pipette mix, leave pipette tip in a tube, pipette mix every ~3–5 min for 15 min.

- **Wash beads:**
  - Magnetically separate, remove the supernatant.
  - Add 1 mL **75% EtOH**, pipette resuspend and move to a **new tube.**
  - After 30 s, magnetically separate, remove the supernatant, and wash 2× with **75% EtOH** (30 s).
  - Magnetically separate, remove residual liquid with fine tip → Air-dry 5 min.
- **Elute RNA:**
  - Resuspend in **10 μL H_2O**, let it sit for 5 min.
  - Magnetically separate, transfer the supernatant to a new tube.

3.9 Reverse Transcribe RNA (All Clip and Input Samples), and Reaction Cleanup

3.9.1 Anneal Primer in 8-Well Strip Tubes

- Mix 10 μL of RNA with 0.5 μL **InvAR17** primer.
- Heat at 65 °C for 2 min in preheated PCR block, place immediately on ice.

3.9.2 Prepare Reverse Transcription Master Mix on Ice; 10 μL Per Sample

– H_2O	4.0 μL
– 10× AffinityScript Buffer	2.0 μL
– 0.1 M DTT	2.0 μL
– dNTPs (**100 mM; 25 mM each**)	0.8 μL
– Murine RNase Inhibitor	0.3 μL
– AffinityScript Enzyme	0.9 μL

- Add 10 μL to each sample, mix, and incubate at 55 °C for 45 min in preheated PCR block.

3.9.3 Cleanup cDNA

- **Removal of excess primers**
  - Add 3.5 μL **ExoSAP-IT** to each sample, vortex, spin down.
  - Incubate at 37 °C for 15 min on a PCR block.
  - Add 1 μL **0.5 M EDTA**, pipette-mix.
- **RNA removal**
  - Add 3 μL **1 M NaOH**, pipette-mix.
  - Incubate at 70 °C for 12 min on a PCR block.
  - Add 3 μL **1 M HCl**, pipette-mix.

Silane Cleanup cDNA

- **Prepare beads:**
  - Magnetically separate 10 μL **MyONE Silane beads** per sample, remove the supernatant.
  - Wash 1× with 500 μL **RLT buffer** and resuspend beads in 93 μL **RLT buffer.**
- **Bind RNA:**
  - Add beads in 93 μL **RLT buffer** to sample, mix and add 111.6 μL **100% EtOH.**
  - Pipette mix, leave pipette tip in a tube, pipette mix twice, for 5 min.
- **Wash beads:**
  - Magnetically separate, remove the supernatant.
  - Add 1 mL **80% EtOH**, pipette resuspend and move to **new tube.**
  - After 30 s, magnetically separate, remove the supernatant, and wash 2× with **80% EtOH** (30 s).
  - Magnetically separate, remove residual liquid with fine tip → air-dry 5 min.
- **Elute RNA:**
  - Resuspend in 5 μL 5 mM Tris–HCl pH 7.5, let it sit for 5 min (do **NOT** remove from beads).

3.10 3′ Linker Ligate cDNA (On-Bead), and Cleanup

3.10.1 Anneal Linker

- Add 0.8 μL **InvRand3Tr3** adapter (*see* **Note 3**).
- Add 1 μL 100% **DMSO.**
- Heat at 75 °C, 2 min, place immediately on ice for >1 min.

3.10.2 Prepare Ligation Master Mix on Ice; 12.8 μL Per Sample

– 10× RNA Ligase Buffer (with DTT)	2.0 μL
– 0.1 M ATP	0.2 μL
– 50% PEG 800	9.0 μL
– High concentration T4 RNA Ligase	0.5 μL
– H_2O	1.1 μL

- Flick to mix, spin down, and add 12.8 μL to each sample: add master mix slowly with stirring; it needs to be homogeneous.
- Add an additional 1 μL **High concentration T4 RNA Ligase on the top of sample** and pipette mix.
- Incubate at room temperature overnight.

3.10.3 Silane Cleanup Linker-Ligated cDNA

- **Prepare beads:**
  - Magnetically separate 5 μL **MyONE Silane beads** per sample, remove the supernatant.

  - Wash 1× with 500 μL **RLT buffer.**
  - Resuspend beads in 60 μL **RLT buffer** per sample.
- **Bind RNA:**
  - Add beads in 60 μL **RLT buffer** to each sample, mix and add 60 μL **100% EtOH**.
  - Pipette mix, incubate for 5 min at room temperature (pipette mix twice with same tips during incubation).
- **Wash beads:**
  - Magnetically separate, remove the supernatant.
  - Add 1 mL **75% EtOH**, pipette resuspend and move to **new tube**.
  - After 30 s, magnetically separate, remove the supernatant.
  - Wash 2× with **75% EtOH** (30 s).
  - Magnetically separate, remove residual liquid with fine tip → Air-dry 5 min.
- **Elute RNA:**
  - Resuspend in 27 μL **10 mM Tris–HCl pH 7.5**, let it sit for 5 min.
  - Magnetically separate, transfer **25 μL** sample to a new tube.

3.11 qPCR Quantify cDNA, PCR, and Reaction Cleanup

3.11.1 Prepare qPCR Master Mix; 9 μL Per Sample

– qPCR 2× master mix	5.0 μL
– H_2O	3.6 μL
– qPCR primer mix	0.4 μL (10 μM each qPCR-grade D5x/D7x mix)

- Mix, dispense into a 384-well qPCR plate.
- Add **1 μL 1:10 diluted (in H_2O) sample cDNA**, seal, mix.
- Run qPCR, note Cq values.

3.11.2 PCR Amplify cDNA

Prepare PCR on ice; 50 μL total per sample:	PCR conditions (cycle # depending on library):
• 2× PCR master mix: 25.0 μL • H_2O: 7.5 μL • 20 μM forward primer (D50x): 2.5 μL • 20 μM reverse primer (D70x): 2.5 μL • Sample cDNA: 12.5 μL	– 98 °C for 30 s – 98 °C for 15 s → 68 °C for 30 s → 72 °C for 40 s (×6 cycles) – 98 °C for 15 s → 72 °C for 60 s (×[qPCR Cq minus 9] cycles) – 72 °C 1 min – 4 °C hold *Total cycle # for final PCR: 3 cycles less than the qPCR Ct or Cq of the 1:10 diluted sample*

- Dispense into 8-well strips, add **12.5 μL CLIP sample + 2.5 μL H_2O** and mix.
- Perform PCR as indicated above.

3.11.3 SPRI Cleanup Library

- Add 90 μL **AmpureXP beads** suspension (do *not* separate) per 50 μL PCR reaction, mix, incubate at room temperature for 10 min (pipette mix 2–3× during incubation).
- Magnetically separate, wash beads 2× with **75% EtOH** remove the supernatant → air-dry beads 5 min.
- Resuspend in **20 μL H_2O**, let it sit for 5 min, magnetically separate for 5 min.
- Transfer **18 μL** to new tubes.

3.12 Agarose Gel Electrophoresis and Size-Selection of Library

- Prepare **3% low-melting temp agarose gel with 1:10,000 SybrSafe** in 1× TBE.

3.12.1 Prepare Samples and Run Gel

- Add 6 μL **6× OrangeG** buffer to each sample (18 μL of sample), mix.
- Load on gel, leave one empty well between samples, **50 bp ladder** on both sides of the gel.
- Run ~95 V for 50 min (longer gives better resolution but larger cut sizes).

3.12.2 Gel-Extract Library from Gel

- Under blue light illumination, cut gel slice from 175–350 bp and place into 15 mL conical tube.
- Use fresh razor blades for each sample and keep cross-contamination to minimum.
- **Cut and elute gel** using Qiagen MinElute gel extraction kit:
  - Add 6× volumes of **Buffer QG** to melt gel (e.g., for 100 mg gel, add 600 μL QG).
  - Melt gel at room temp (do *not* heat) on benchtop (can shake to help melt, but do not vortex).
  - Add 1× volume of original gel of **isopropanol** and *mix well* (100 mg gel = 100 μL isopropanol).
  - Load on column (750 μL per spin, can do multiple spins, all spins max speed 1 min).
  - After all sample has been spun through, wash 1× with 500 μL **Buffer QG.**
  - Add 1× with 750 μL **Buffer PE**, spin 1 min, pour out flow-through, spin again 2 min max speed.
  - Carefully move column to a new 1.5 mL tube, and air dry for 2 min.
  - Carefully add 12.5 μL **Buffer EB** directly to the center of the column, incubate for 2 min at room temperature, spin at max speed.

3.13 Quantitate Library

- Quantitate using D1000 DNA Tapes for the Agilent TapeStation.

4 Notes

1. **RNase inhibitor addition to lysis buffer (Subheading 3.2.1).**

   Murine RNase inhibitor (NEB) inhibits many endogenous RNase enzymes but does not significantly inhibit RNase I. As such, it can be added either before or after RNase I treatment. For many cell lines (HEK293T, K562) we have observed that this choice does not alter fragmentation or ultimate signal. However, we have observed that for cell types or tissues with moderate to high endogenous RNase activity (e.g., stem cells, differentiated neurons, many tissues), the addition of RNase inhibitor at the initial lysis step is essential to prevent over-fragmentation. We note that the amount of RNase inhibitor may need to be further increased for samples with particularly high RNase activity (e.g., liver or pancreas).

2. **Buffer transitions at wash (Subheadings 3.4.1, 3.4.4, and others).**

   We often perform intermediate washes with equal mixtures of the previous and current wash buffer, in order to decrease dramatic changes in buffer composition.

3. **Modified adaptor strategy for single-end enhanced CLIP (seCLIP) (Subheadings 3.5.1, 3.8.1, and 3.10.1).**

   eCLIP and iCLIP methods utilize the tendency of reverse transcriptase to terminate at the protein-crosslinked RNA nucleotide to enable single-nucleotide resolution of binding sites. In the initial eCLIP adaptor strategy (referred to here as paired-end peCLIP), this position was located at the beginning of the second (paired-end) read in standard Illumina sequencing, which yielded high-quality data but added additional cost. To enable single-end eCLIP (seCLIP), we created a modified adapter strategy that inverted this read structure (Fig. 2a), but used the same highly efficient enzymatic steps (including a 3′ RNA adapter ligation to the RNA fragments, reverse transcription using a melting temperature-optimized primer, and a 3′ ssDNA ligation to the cDNA fragment). Performing both peCLIP and seCLIP on well-characterized splicing regulator RBFOX2 in HEK293T cells, we observed similar read density profiles for individual example binding sites (Fig. 2b) and comparable results in library yield (quantified as the number of PCR cycles necessary to obtain 100 femtomoles of library (eCT)) with biological replicates averaging 12.3 and 12.7 eCT for peCLIP and seCLIP respectively (Fig. 2c). Transcriptome-wide, we observed high correlation of read density within peaks between peCLIP and seCLIP ($R^2 = 0.46$ and 0.52), equivalent to those observed between bio-

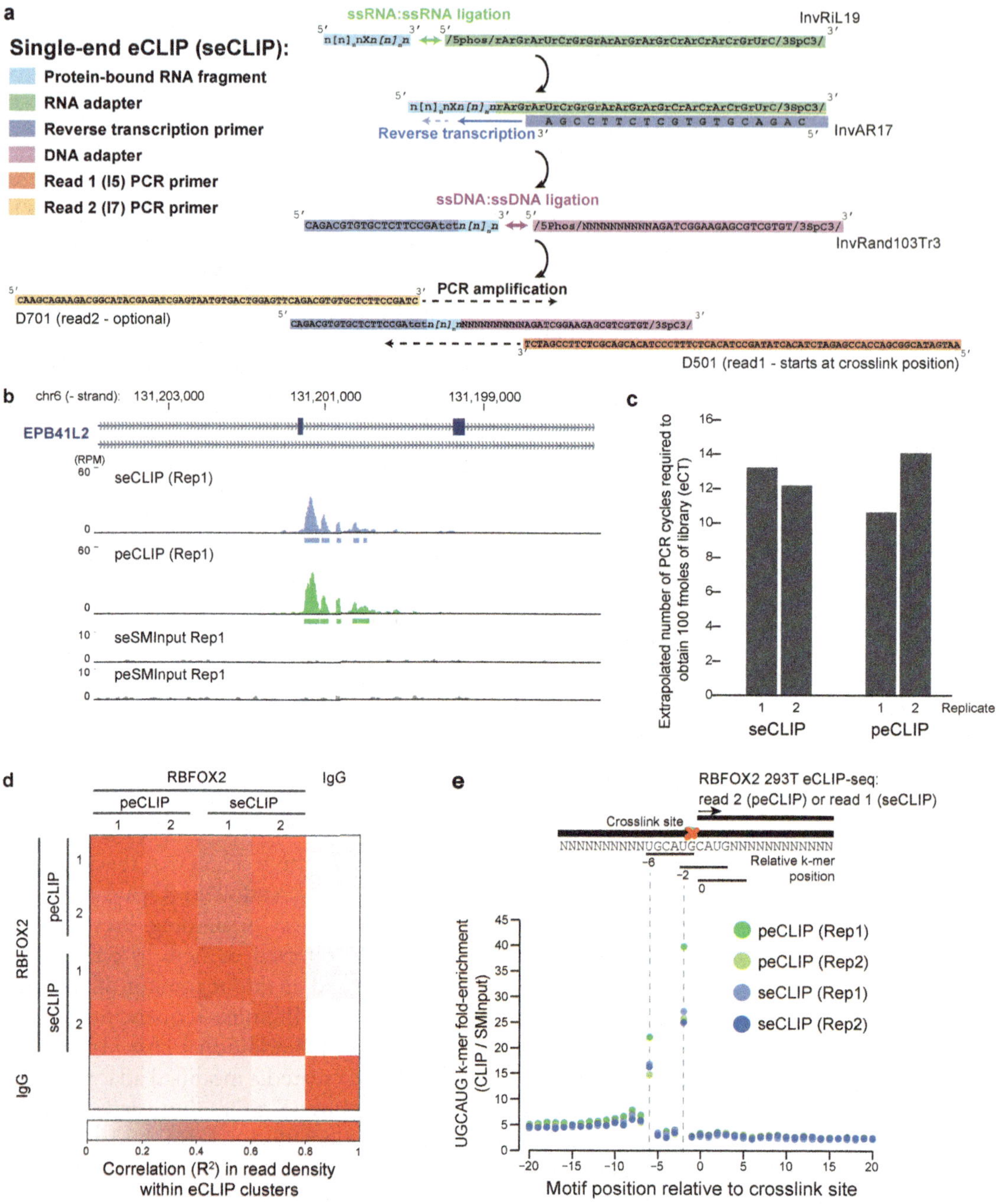

Fig. 2 Modified adaptor strategy for single-end enhanced CLIP (seCLIP). (**a**) Schematic of adaptor sequences used in seCLIP. (**b**) Read density (shown as reads per million; RPM) observed in original, paired-end eCLIP (peCLIP) and seCLIP for RBFOX2 in HEK293T cells at EPB41L2 exon 13-14. Boxes below tracks indicate significantly enriched peaks after input normalization. (**c**) Required amplification observed using seCLIP and peCLIP adaptor strategies for two biological replicates profiling RBFOX2 in HEK293T. (**d**) Heatmap indicates correlation across experiments for read fold-enrichment in CLIP versus input, considering peaks identified in indicated experiments (y-axis). (**e**) Plot indicates enrichment for RBFOX2 binding motif UGCAUG at indicated positions around read start positions

logical replicates of peCLIP (R^2 = 0.52) or seCLIP (R^2 = 0.47) (Fig. 2d). Confirming that seCLIP maintains single-nucleotide resolution, we observed the same stereotypical enrichment for the RBFOX family motif (UGCAUG) enriched at the −6 and −2 positions relative to read start positions (Fig. 2e). Thus, this modified seCLIP adapter strategy enables the same high-quality eCLIP data generation amenable to sequencing with standard single-end, 50 bp chemistry, decreasing cost. Further, the use of standard Illumina barcodes in seCLIP enables pooling with other standard Illumina RNA-seq or other high-throughput sequencing libraries.

4. **Decreasing contamination introduced by nitrocellulose membrane sources (Subheading 3.6.4).**

In 102 K562 and HepG2 eCLIP experiments, an average of 84% of reads uniquely or multiply mapped to the human genome, respectively [11]. However, in preliminary experiments in cell types with decreased RNA yield after membrane transfer (motor neurons (MN) and neural progenitor cells (NPCs) derived from human embryonic stem cells), we observed in some cases more than 90% of sequenced reads were not mapped to the human genome. Using SOAP-denovo [13] to de novo assemble the unmapped reads, we assembled multiple contigs that were queried against the NR database and showed >99% identity to *Acinetobacter johnsonii XBB1* (CP010350.1). Re-mapping these eCLIP datasets revealed millions of reads in many datasets mapping throughout CP010350.1, confirming this specific species as a major contamination source (Fig. 3a). We noted that reads mapping to CP010350.1 had proper CLIP adapter structure, indicating that contamination was likely occurring prior to the 3′ linker ligation.

In order to modify eCLIP to ameliorate this issue, we set out to identify the source of this contamination by performing RT-qPCR using primers designed against regions of CP010350.1 with high read density. To first confirm that this contamination was not present in initial samples, we extracted RNA (Trizol LS) from supernatant remaining after immunoprecipitation (Subheading 3.4.1) in addition to standard post-membrane transfer and isolation (Subheading 3.7.3) and performed RT-qPCR for bacterial RNA. We observed more than tenfold increased bacterial RNA signal in the membrane-isolated RNA as compared to supernatant RNA from the equivalent number of cells, indicating that the contamination was not present during tissue culture and was introduced during the IP/Western stage (Fig. 3b). Similar RT-qPCR assays performed after RNA isolation on various buffers or enzyme mixes used failed to identify significant contamination (data not shown).

The observation that these reads were only present in input but not CLIP samples (despite input often having more

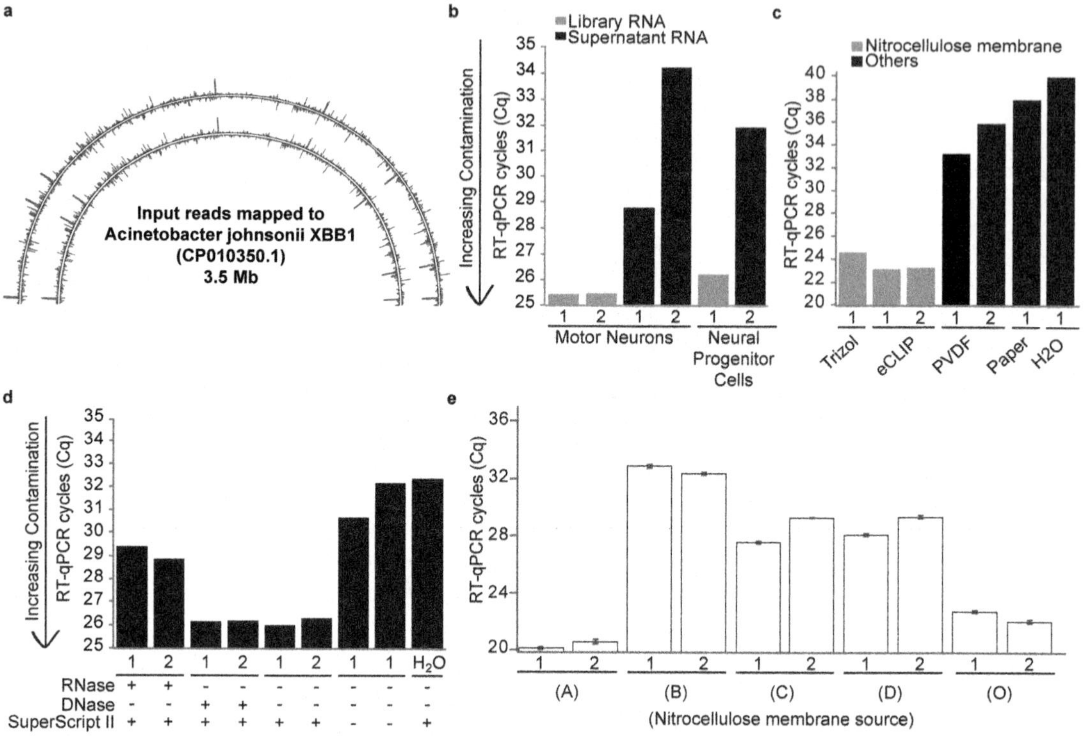

Fig. 3 Genome of Acinetobacter johnsonii XBB1 in nitrocellulose membranes detected by RT—qPCR. (**a**) Sequencing reads from two eCLIP input libraries mapped to Acinetobacter johnsonii XBB1 (CP010350.1). (**b**) Bars indicate Cq from RT—qPCR performed using CP010350.1-specific qPCR primers on eCLIP RNA (Subheading 3.7.3) and supernatant samples (Subheading 3.4.1) from indicated cell types. Lower Cq reflects higher CP010350.1 signal. Numbers indicate replicate experiments. (**c**) Bars indicate RT—qPCR Cq for CP010350.1 from RNA isolated from nitrocellulose membranes via indicated method, PVDF membranes (RNA isolation by Trizol), paper (RNA isolation by Trizol), and H_2O (Trizol extraction). (**d**) Bars indicate RT—qPCR Ct for CP010350.1 from RNA extracted from size-matched nitrocellulose membranes, in technical replicates. Symbols below indicate samples that were either RNase or DNase treated, and those with and without RT enzyme added. (**e**) Bars indicate RT—qPCR Ct for CP010350.1 from RNA extracted from nitrocellulose membrane samples from five sources in technical replicates as follows: (A) commercial source A, (B) ThermoFisher iBlot (IB23001 lot 2NR26016-01), (C) GE Amersham Protran Premium (13600117 lot G6552142), (D) GE Amersham Protran Premium (1060008 lot G9931040), and (O) original commercial source. Error bars indicate standard deviation from RT—qPCR triplicate measurements

than 100-fold more RNA recovery and library yield [12]), implicated the difference in RNA adaptor ligation of CLIP (3′ RNA adaptor ligation on-bead, before the protein electrophoresis step) versus input (3′ RNA adaptor ligation after RNA isolation off of membranes) RNA. Surprisingly, we found that RNA extraction (Trizol) of nitrocellulose membrane alone yielded RT-qPCR signal similar to our contaminated libraries, in contrast to the lower signal observed after RNA isolation from PVDF membranes, Whatman and other lab paper, or negative controls (Fig. 3c). We observed identical results with freshly ordered membrane stock (data not shown). To further explore the nature of the contamination, we synthesized cDNA from either RNase or DNase-treated membrane samples and repeated the RT-qPCR assay. This indicated that the contamination was likely RNA, as the RT-qPCR signal was sensitive to both RNase and the no-RT control, but not DNase (Fig. 3d). The strand-specific signal observed in reads similarly implicated RNA contamination (Fig. 3a).

As the nitrocellulose transfer provides key specificity for isolating RNA crosslinked to protein, we set out to identify optimized alternative sources that had decreased RNA background. We obtained four additional nitrocellulose membrane sources (A), (B) ThermoFisher iBlot2 (IB23001 lot 2NR26016-01), (C) GE Amersham Protran Premium (13600117 lot G6552142), and (D) GE Amersham Protran Premium (1060008 lot G9931040), in addition to our original commercial source (O). For each, we performed RNA isolation followed by the bacterial RNA RT-qPCR, and observed that whereas O and A showed similar CP010350.1 contamination, B, C, and D did not (Fig. 3e). When we prepared libraries according to our standard protocol for eCLIP input samples, we observed that library yields reflected these results, with B, C, and D showing the least amount of overall contamination (data not shown). Importantly, we observed no difference in RNA or library yield when we prepared standard eCLIP input libraries for two protein size ranges of roughly equal membrane size (10–50 kDa and 50–225 kDa) for multiple membrane types. Thus, these results indicate that testing of nitrocellulose membranes enables optimization of eCLIP by removing substantial background contamination. Our results identify (B) ThermoFisher iBlot and (C) GE Amersham Protran Premium (1060008) membranes as options which show a dramatic decrease in contamination for sensitive eCLIP experiments without altering true library yield. Although yielding equally low contamination, (D) GE Amersham Protran Premium (13600117) is "trial-size" packaging and generally commercially unavailable for large-scale use. Other sources can be tested using the RT-qPCR method described here to determine whether they are of sufficiently low background for use in eCLIP.

Acknowledgments

The authors would like to thank members of the Yeo lab for insightful discussions and critical reading of the manuscript, particularly S. Aigner. This work was supported by grants from the National Institute of Health [HG004659, HG007005, and NS075449 to G.W.Y.]. E.L.V.N. is a Merck Fellow of the Damon Runyon Cancer Research Foundation [DRG-2172-13]. G.A.P. is supported by the National Science Foundation Graduate Research Fellowship. G.W.Y. is an Alfred P. Sloan Research Fellow.

References

1. Morris KV, Mattick JS (2014) The rise of regulatory RNA. Nat Rev Genet 15(6):423–437. doi:10.1038/nrg3722
2. Quinn JJ, Chang HY (2016) Unique features of long non-coding RNA biogenesis and function. Nat Rev Genet 17(1):47–62. doi:10.1038/nrg.2015.10
3. Jonas S, Izaurralde E (2015) Towards a molecular understanding of microRNA-mediated gene silencing. Nat Rev Genet 16(7):421–433. doi:10.1038/nrg3965
4. Glisovic T, Bachorik JL, Yong J, Dreyfuss G (2008) RNA-binding proteins and post-transcriptional gene regulation. FEBS Lett 582(14):1977–1986. doi:10.1016/j.febslet.2008.03.004
5. Gerstberger S, Hafner M, Tuschl T (2014) A census of human RNA-binding proteins. Nat Rev Genet 15(12):829–845. doi:10.1038/nrg3813
6. Licatalosi DD, Darnell RB (2010) RNA processing and its regulation: global insights into biological networks. Nat Rev Genet 11(1):75–87. doi:10.1038/nrg2673
7. Konig J, Zarnack K, Luscombe NM, Ule J (2011) Protein-RNA interactions: new genomic technologies and perspectives. Nat Rev Genet 13(2):77–83. doi:10.1038/nrg3141
8. Ule J, Jensen KB, Ruggiu M, Mele A, Ule A, Darnell RB (2003) CLIP identifies Nova-regulated RNA networks in the brain. Science 302(5648):1212–1215. doi:10.1126/science.1090095
9. Hafner M, Landthaler M, Burger L, Khorshid M, Hausser J, Berninger P, Rothballer A, Ascano M Jr, Jungkamp AC, Munschauer M, Ulrich A, Wardle GS, Dewell S, Zavolan M, Tuschl T (2010) Transcriptome-wide identification of RNA-binding protein and microRNA target sites by PAR-CLIP. Cell 141(1):129–141. doi:10.1016/j.cell.2010.03.009
10. Zhang C, Darnell RB (2011) Mapping in vivo protein-RNA interactions at single-nucleotide resolution from HITS-CLIP data. Nat Biotechnol 29(7):607–614. doi:10.1038/nbt.1873
11. Konig J, Zarnack K, Rot G, Curk T, Kayikci M, Zupan B, Turner DJ, Luscombe NM, Ule J (2010) iCLIP reveals the function of hnRNP particles in splicing at individual nucleotide resolution. Nat Struct Mol Biol 17(7):909–915. doi:10.1038/nsmb.1838
12. Van Nostrand EL, Pratt GA, Shishkin AA, Gelboin-Burkhart C, Fang MY, Sundararaman B, Blue SM, Nguyen TB, Surka C, Elkins K, Stanton R, Rigo F, Guttman M, Yeo GW (2016) Robust transcriptome-wide discovery of RNA-binding protein binding sites with enhanced CLIP (eCLIP). Nat Methods 13(6):508–514. doi:10.1038/nmeth.3810
13. Luo R, Liu B, Xie Y, Li Z, Huang W, Yuan J, He G, Chen Y, Pan Q, Liu Y, Tang J, Wu G, Zhang H, Shi Y, Liu Y, Yu C, Wang B, Lu Y, Han C, Cheung DW, Yiu SM, Peng S, Xiaoqian Z, Liu G, Liao X, Li Y, Yang H, Wang J, Lam TW, Wang J (2012) SOAPdenovo2: an empirically improved memory-efficient short-read de novo assembler. GigaScience 1(1):18. doi:10.1186/2047-217X-1-18

Chapter 15

Purification of Transcript-Specific mRNP Complexes Formed In Vivo from *Saccharomyces cerevisiae*

Jenna E. Smith and Kristian E. Baker

Abstract

RNA binding proteins play critical roles in shaping the complex life cycle of cellular transcripts. For most RNAs, the association with a distinct complement of proteins serves to orchestrate its unique pattern of maturation, localization, translation, and stability. A key aspect to understanding how transcripts are differentially regulated lies, therefore, in the ability to identify the particular repertoire of protein binding partners associated with an individual transcript. We describe here an optimized experimental procedure for purifying a single mRNA population from yeast cells for the characterization of transcript-specific mRNA-protein complexes (mRNPs) as they exist in vivo. Chemical cross-linking is used to trap native mRNPs and facilitate the co-purification of protein complexes associated with an individual transcript population that is captured under stringent conditions from cell lysates through hybridization to complementary DNA oligonucleotides. The resulting mRNP is highly enriched and largely devoid of non-target transcripts, and can be used for a number of downstream analyses including protein identification by mass spectrometry.

Key words RNA, RNA binding proteins, mRNA, mRNP, Affinity capture, Mass spectrometry, Yeast

1 Introduction

The posttranscriptional fate of an mRNA is strongly influenced by its protein binding partners. Indeed, RNA binding proteins (RBPs) play critical roles in mediating and regulating the processing, transport, localization, translation, and decay of all cellular protein-coding transcripts [1, 2]. Given their importance, significant efforts have been made over the years to experimentally identify RBPs and their targets, and to understand the physiological function of mRNA-protein complexes (mRNPs) in the life cycle of RNA.

In general, two approaches have been employed to isolate RNA-protein complexes (reviewed in [3, 4]). The first uses immunoprecipitation to collect a specific RBP of interest from cell lysates and the RNAs that co-purify with it. In this method, it is advantageous to covalently cross-link RNA-protein complexes (by treating intact cells with either formaldehyde or UV light) to increase the

Yongsheng Shi (ed.), *mRNA Processing: Methods and Protocols*, Methods in Molecular Biology, vol. 1648,
DOI 10.1007/978-1-4939-7204-3_15, © Springer Science+Business Media LLC 2017

capture of proteins that associate with the target RNA only transiently and to minimize protein re-association during the purification that results in interactions that do not reflect bona fide in vivo mRNPs [5, 6]. When coupled with next-generation sequencing technologies, this approach can be used to identify targets of a known RBP at the level of the entire transcriptome and map the binding position along the RNA at near-nucleotide resolution [7, 8]. In the second method, referred to generally as RNA capture, an RNA of interest is targeted and retrieved from cell lysates for the purpose of co-purifying and characterizing the novel RBPs that interact with it. For the isolation of RNA-protein complexes that exist in vivo, the target RNA is typically expressed as a transgene harboring sequence tags within its 3′ UTR that bind with high affinity to either a protein (e.g., bacteriophage MS2 coat protein) or small molecule (e.g., tobramycin or streptavidin) that can be conjugated to a resin or magnetic bead [9–13]. The global and unbiased identification of proteins that co-purify with the target RNA can then be achieved using quantitative or nonquantitative mass spectrometry.

Recent advances in RNA capture methodologies have employed hybridization to complementary oligonucleotides as a means to affinity capture endogenous RNA targets and circumvent the need to ectopically express tagged transgenes. In addition, a cross-linking step is often added to allow purification under denaturing conditions so as to isolate native interactions and enhance specificity in the RNA capture. Using these additions, total cellular polyA$^+$ RNA and its directly associated proteins have been purified through hybridization to oligo-d(T) for the systematic identification of mRNA-interacting proteins in a number of cell types [14–20]. In addition, mammalian long noncoding RNA *Xist* has been captured using antisense oligonucleotides to reveal novel protein interacting partners critical for its function in transcriptional silencing of the X chromosome [21, 22]. These studies clearly demonstrate the utility of purifying RNA-protein complexes that exist on distinct transcript populations as a means to unravel the function or events that contribute to the unique cellular lifetime of the RNA.

Here, we describe the application of RNA affinity capture in the purification of a single mRNA population from yeast cell lysates for the isolation of a transcript-specific mRNP complex as it exists in vivo. Importantly, growing cells are treated in situ with limiting formaldehyde to mediate formation of chemical cross-links between protein-RNA and protein-protein complexes as they are assembled in vivo [23]. This treatment traps highly dynamic and transient interactions that exist between proteins and the RNA target, enables the capture of proteins that associate both directly and indirectly (such as those from multi-subunit protein complexes) with the transcript, and allows for purification of native mRNP complexes under highly stringent and denaturing conditions.

Target RNA is purified from whole cell lysates through hybridization to DNA oligonucleotides with complementarity to various regions along the transcript and which are biotinylated and immobilized to streptavidin-conjugated magnetic beads. Purified mRNPs are highly enriched for the RNA target (~10,000-fold, as demonstrated by northern blot analysis and quantitative RT-PCR) and protein components can be identified either individually by SDS-PAGE and western blot or globally by mass spectrometry (Fig. 1). We recently utilized this approach to identify proteins that associate with an mRNA targeted to nonsense-mediated mRNA decay

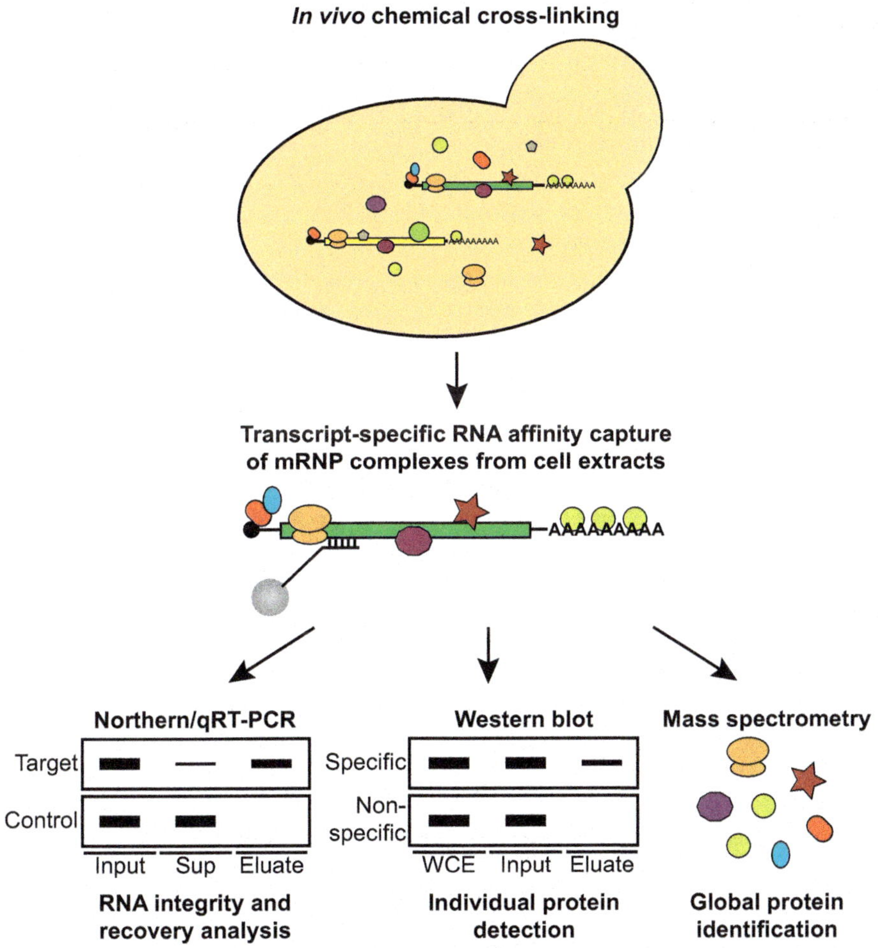

Fig. 1 Schematic of transcript-specific mRNP affinity capture. Native protein-protein and protein-nucleic acid interactions are stabilized in vivo by covalent cross-linking with the addition of formaldehyde to growing yeast cells. Cells are harvested and lysed, and the extracts incubated under denaturing, stringent conditions with magnetic bead-bound DNA oligonucleotides complementary to a single RNA population. Affinity captured, transcript-specific mRNPs are washed with denaturing buffers and low salt and processed for downstream applications including targeted RNA or protein analysis to monitor mRNP integrity and specificity of capture, or global protein identification by mass spectrometry

(NMD) compared to a matched NMD-insensitive mRNA [24]. We found ~100 proteins associated with the control transcript—a majority of which are polyA$^+$ RNA binding proteins defined in recent global studies [16, 18]. Critically, we found ~50 proteins enriched exclusively on the NMD target, including UPF1, a key component of the NMD machinery [24]. This method offers a facile approach to purify transcript-specific mRNP complexes from yeast cells for the identification and characterization of RBPs that contribute to regulating the cellular fate of that RNA.

2 Materials

(*see* **Note 1**).

2.1 Cell Growth, Chemical Cross-Linking, and Harvesting

1. Synthetic Dextrose (SD) media with appropriate amino acids supplemented: 0.17% Bacto-Yeast Nitrogen Base, 0.5% ammonium sulfate, 2% dextrose, 0.2% Amino Acid Dropout Mix (*see* **item 2** below), pH to 6.5 with 10 N NaOH. Sterilized by autoclave.
2. Amino Acid Dropout Mix: Combine the following powder reagents: 0.25 g adenine: 1 g alanine: 1 g arginine: 1 g asparagine: 1 g aspartic acid: 1 g cysteine: 1 g glutamine: 1 g glutamic acid: 1 g glycine: 1 g inositol: 1 g isoleucine: 1 g lysine: 1 g methionine: 1 g para-aminobenzoic acid: 1 g phenylalanine: 1 g proline: 1 g serine: 1 g threonine: 1 g tyrosine: 1 g valine: 1 g uracil: 1 g histidine: 1 g tryptophan: 5 g leucine. Omit amino acid(s) as required for the maintenance of a plasmid-borne auxotrophic marker (*see* **Note 2**). Mix thoroughly using a mortar and pestle.
3. Appropriate yeast strain (*see* **Note 3**).
4. 30 °C shaker/incubator.
5. Visible light spectrophotometer capable of reading at 600 nm λ and cuvettes.
6. Formaldehyde (37% stock).
7. 2.5 M glycine. Sterilized by autoclave.
8. Clinical centrifuge (or similar) capable of processing sample volumes of 200 mL at speeds of 3200 × *g*.
9. 2.0 mL Eppendorf tubes.
10. Microcentrifuge capable of processing 2.0 and 1.5 mL Eppendorf tubes at speeds of up to 16,100 × *g*.
11. Dry ice.
12. −80 °C freezer (for storage of samples [cell pellets, lysates, RNA, or protein]).

2.2 Cell Lysis

1. RNase-free dH_2O (*see* **Note 4**).
2. 1× Lysis Buffer: 10 mM Tris–HCl pH 7.4, 100 mM NaCl, 30 mM $MgCl_2$, protease inhibitor. Prepared using RNase-free dH_2O.
3. 0.5 M dithiothreitol (DTT).
4. Sterile glass beads: 0.5 mm diameter.
5. Vortexer housed at 4 °C.
6. 18 gauge needles.
7. Bunsen burner or similar heat source.
8. 15 mL conical bottom centrifuge tubes.
9. Clinical centrifuge (or similar) capable of processing 15 mL conical bottom centrifuge tubes at 1600 × *g*, housed at 4 °C.
10. 1.5 mL Eppendorf tubes.
11. 2.0 mL Eppendorf tubes.
12. Microcentrifuge capable of processing 2.0 and 1.5 mL Eppendorf tubes at speeds of up to 16,100 × *g*.
13. UV spectrophotometer and quartz cuvettes, capable of reading at OD_{260} and OD_{280} (optional).

2.3 Magnetic Bead Preparation

1. MyOne Streptavidin C1 Dynabeads and magnet (*see* **Note 5**).
2. 1.5 mL Eppendorf tubes.
3. 1× Binding and Washing Buffer (1× B&W Buffer): 5 mM Tris–HCl pH 7.5, 0.5 mM EDTA pH 8.0, 1 M NaCl. Prepared using RNase-free dH_2O.
4. Biotinylated oligonucleotides (*see* **Note 6**): custom sequence-specific DNA oligonucleotides, with a 3′ biotinylation modification. Resuspended to 400 pmol/μL using RNase-free dH_2O.
5. 2× Binding and Washing Buffer (2× B&W Buffer): 10 mM Tris–HCl pH 7.5, 1 mM EDTA pH 8.0, 2 M NaCl. Prepared using RNase-free dH_2O.
6. Tube rotator capable of end-over-end rotation.

2.4 Hybridization/Affinity Capture

1. TE Buffer: 10 mM Tris–HCl, 1 mM EDTA, pH to 8.0. Sterilized by autoclave.
2. 15 mL conical bottom centrifuge tubes.
3. 5 M LiCl. Sterilized by autoclave.
4. 10% SDS. Filter sterilized.
5. 0.5 M EDTA pH 8.0. Sterilized by autoclave.
6. M Tris–HCl pH 7.5. Sterilized by autoclave.
7. 100% Formamide.

8. Fungal protease inhibitor.
9. Tube rotator capable of end-over-end rotation.
10. 1.5 mL Eppendorf tubes.
11. Wash Buffer 1: 10 mM Tris–HCl pH 7.5, 1 mM EDTA pH 8.0, 250 mM LiCl, 0.1% SDS. Prepared using RNase-free dH_2O.
12. Wash Buffer 2: 10 mM Tris–HCl pH 7.5, 1 mM EDTA pH 8.0, 100 mM LiCl. Prepared using RNase-free dH_2O.
13. Microcentrifuge capable of processing 2.0 and 1.5 mL Eppendorf tubes at speeds of up to 16,100 × *g*.

2.5 Elution of Purified mRNP Material

1. RNase-free dH_2O.
2. 70 °C heat block.
3. 1.5 mL Eppendorf tubes.

2.6 RNA Analysis (Optional, But Recommended)

1. 5 M NaCl. Sterilized by autoclave.
2. GlycoBlue nucleic acid coprecipitant.
3. 95% and 75% ethanol. Stored at −20 °C.
4. Microcentrifuge capable of processing 1.5 mL Eppendorf tubes at speeds of up to 16,100 × *g*.
5. LET Buffer: 25 mM Tris–HCl pH 8.0, 100 mM LiCl, 20 mM EDTA pH 8.0. Sterilized by autoclave.
6. 10% SDS. Filter sterilized.
7. 65 °C heat block.
8. P/C/L: 50% v/v phenol/50% v/v chloroform, equilibrated with LET Buffer.
9. Vortexer housed at room temperature.
10. 1.5 mL Eppendorf tubes.
11. Chloroform.
12. RNase-free dH_2O.

2.7 Protein Analysis by SDS-PAGE (Optional)

1. Vacuum concentrator.
2. 100% and 80% acetone. Stored at −20 °C.
3. Microcentrifuge capable of processing 2.0 and 1.5 mL Eppendorf tubes at speeds of up to 16,100 × *g*, housed at 4 °C.
4. 1× SDS Sample Buffer: 125 mM Tris–HCl pH 6.8, 2% SDS, 100 mM DTT, 10% glycerol, 0.05% bromophenol blue.
5. 70 °C heat block.
6. 95 °C heat block.

3 Methods

3.1 Cell Growth, Chemical Cross-Linking, and Harvesting

1. Grow a 200 mL culture of cells (*see* **Notes 3** and 7) at 30 °C and 250 rpm in SD media supplemented with appropriate amino acids (*see* **Note 2**) to an optical density of 0.4 at 600 nm λ (OD_{600}).
2. Cross-link RNA-protein and protein-protein complexes by adding formaldehyde to cell culture at 0.25% (final concentration) and incubating for 15 min at 30 °C and 250 rpm (*see* **Note 8**).
3. Quench cross-linking reaction by adding glycine to the culture to 0.125 M (final concentration) and incubating for 5 min at 30 °C and 250 rpm.
4. Harvest cells by centrifugation in clinical centrifuge at 3200 × *g* for 1 min at room temperature. Decant media from pelleted cells.
5. Resuspend cell pellet in residual media and transfer suspension to a 2 mL Eppendorf tube. Pellet cells at 16,100 × *g* for 30 s to collect sample at bottom of tube, and remove residual supernatant. Flash freeze pelleted cells on dry ice and store at −80 °C.

3.2 Cell Lysis

1. Supplement ice-cold (*see* **Note 9**) 1× Lysis Buffer with fresh 0.5 M DTT to a final concentration of 1 mM.
2. Resuspend each cell pellet in 400 μL 1× Lysis Buffer by vortexing (*see* **Note 10**). Immediately place resuspended cells on ice.
3. Add ½ volume (~200 μL volume) of sterile glass beads to resuspended cells.
4. Vortex for 3 min at 4 °C and then return sample to ice for 2 min. Repeat two additional times.
5. Puncture the bottom of the 2 mL Eppendorf tube with a red-hot 18-gauge needle (heated using a Bunsen burner or other heat sources) and nest tube within the opening of a 15 mL conical bottom centrifuge tube that has been prechilled by placing in ice. Centrifuge at 1600 × *g* for 2 min at 4 °C (*see* **Note 11**).
6. Transfer the supernatant to a cold 1.5 mL Eppendorf tube and adjust volume for any lost material by bringing up volume to 400 μL with 1× Lysis Buffer. Place sample on ice.
7. Resuspend pelleted cell debris from the previous step in 0.5 mL 1× Lysis Buffer and transfer to a 2 mL Eppendorf tube that has been prechilled by placing in ice. Pellet at 16,100 × *g* for 30 s, and remove the residual supernatant. Repeat lysis (Subheading 3.2, **steps 2–6**) and combine supernatant with lysate from first lysis (*see* **Note 12**). The total lysate volume should now equal

approximately 800 μL. Store yeast whole cell lysate on ice until steps in Subheading 3.4.

8. Optional: quantify lysate by measuring the OD_{260} and OD_{280} with a UV spectrophotometer using quartz cuvettes and diluting 5 μL in 995 μL dH_2O (*see* **Note 13**). Use an equivalent dilution of 1× Lysis Buffer in dH_2O as a blank.

3.3 Magnetic Bead Preparation

1. For each lysed 200 mL cell pellet, resuspend MyOne Streptavidin C1 Dynabeads completely by repeat pipetting, and transfer 250 μL beads to a 1.5 mL Eppendorf tube (*see* **Note 5**).
2. Place tube alongside magnet and allow beads to collect along side of tube. Remove the supernatant by pipetting.
3. Wash beads with an equal volume (250 μL) of 1× B&W Buffer (*see* **Note 14**) by pipetting up and down approximately 20 times.
4. Place tube alongside magnet and allow beads to collect along side of tube. Remove the supernatant by pipetting.
5. Repeat wash (**steps 3** and **4** above) two more times.
6. Immobilize biotinylated DNA oligonucleotides (*see* **Note 6**) to washed Dynabeads in an incubation volume 4× the original bead volume used. For a 250 μL starting volume of magnetic beads, add reagents as indicated below. Incubate with end-over-end rotation for 15 min at room temperature.

Reagent	Volume
2× B&W Buffer	500 μL
Biotinylated DNA oligonucleotides (400 pmol/μL; *see* **Note 15**)	25 μL
RNase-free dH_2O	475 μL

7. Place tube alongside magnet and allow beads to collect along side of tube. Remove supernatant by pipetting. Wash beads with 500 μL 1× B&W Buffer by pipetting up and down approximately 20 times. Repeat wash step with an additional 500 μL 1× B&W Buffer.
8. Resuspend beads in 250 μL 1× B&W Buffer. Set beads aside at room temperature until ready for use, or continue immediately to the next step.

3.4 Hybridization/Affinity Capture

1. Bring volume of yeast whole cell lysate (800 μL total volume from the two extractions of a cell pellet from 200 mL of culture, from Subheading 3.2, **step 7**) up to 3.5 mL with 2.7 mL TE Buffer. Remove an aliquot as "input" and store at −80 °C—ideal amounts are 1/10 volume for downstream RNA analysis or 1/20 for protein analysis (*see* **Note 16**).

2. Prepare 5 mL (*see* **Note 17**) Hybridization Reaction for each sample in a 15 mL conical tube as follows (*see* **Note 18**).

Reagent	Volume	Final concentration
Yeast lysate + TE buffer	3 mL	N/A
5 M LiCl	500 μL	500 mM LiCl
10% SDS	250 μL	0.5% SDS
0.5 M EDTA pH 8.0	500 μL	50 mM EDTA pH 8.0
1 M Tris–HCl pH 7.5	50 μL	10 mM Tris–HCl pH 7.5
100% formamide	700 μL	14% formamide
Fungal protease inhibitor	5 μL	N/A

3. Anneal RNA to bead-bound DNA oligonucleotides: place Dynabeads from Subheading 3.3, **step 8** alongside magnet, allow beads to collect along side of tube, and remove supernatant. Use a 500 μL aliquot of Hybridization Reaction to resuspend Dynabeads. Transfer aliquot and resuspended beads back to the 15 mL conical bottom centrifuge tube containing the remaining hybridization reaction. Close and seal lid with parafilm and incubate overnight at room temperature with end-over-end rotation (*see* **Notes 19** and **20**).
4. Centrifuge 15 mL conical tube for 5 s at 1600 × *g* to collect sample at bottom of tube.
5. Transfer 1 mL of the Hybridization Reaction/Dynabeads mixture to a 1.5 mL Eppendorf tube. Place tube alongside magnet and allow beads to collect along side of tube; remove the supernatant. Continue adding ~1 mL aliquots of Hybridization Reaction/Dynabeads mixture to the same 1.5 mL Eppendorf tube until the entire Hybridization Reaction sample is transferred and all the beads are collected into one Eppendorf tube. Save an aliquot of the supernatant (i.e., "supernatant") for downstream analysis (e.g., to determine the efficiency of target RNA depletion)—ideally 1/10 volume for RNA analysis or 1/20 volume for protein analysis (*see* **Note 16**).
6. Wash beads twice with 500 μL Wash Buffer 1 (*see* **Note 21**). For each wash, thoroughly resuspend beads using a micropipette, wash by pipetting up and down approximately 20 times, then place tube alongside magnet and allow beads to collect along side of tube. Remove the supernatant and discard.
7. Wash beads three times with 500 μL Wash Buffer 2 (*see* **Note 22**). Perform washes as in Subheading 3.4, **step 6**. Following removal of supernatant after each wash, pulse-spin sample for 10 s in a microcentrifuge, place tube alongside magnet and

allow beads to collect along side of tube. Remove residual supernatant and discard.

8. For RNA analysis or protein analysis by western blot of the entire mRNP sample, proceed to steps in Subheading 3.5. For protein analysis by mass spectrometry, proceed to steps in Subheading 3.8.

3.5 Elution of Purified mRNP Material (See Note 23)

1. Resuspend beads in 75 μL dH_2O per 100 μL beads (i.e., 187.5 μL dH_2O for 250 μL beads).
2. Heat beads at 70 °C for 2 min to disrupt DNA:RNA hybrid annealing interaction (*see* **Note 24**).
3. Place beads immediately alongside magnet and allow beads to collect along side of tube. Transfer eluate to a new 1.5 mL Eppendorf tube and save for downstream processing. In the event that any beads are carried over into the eluate, place the new tube on a magnet and transfer eluate one additional time to a new 1.5 mL Eppendorf tube to ensure complete removal of beads.
4. Remove an aliquot of eluate for RNA recovery analysis (ideally 1/10 volume for RNA; *see* **Notes 16** and **25**). This step may be omitted if RNA analysis is performed on the entire eluate (as in the case of optimization experiments).
5. Proceed to steps in Subheading 3.6 for processing samples for RNA analysis, and steps in Subheading 3.7 for protein analysis.

3.6 RNA Analysis (Optional But Recommended) (See Notes 16, 23, and 26)

1. Bring the eluate volume from steps in Subheading 3.5 up to 400 μL with RNase-free water. Add 5 M NaCl to 0.2 M final concentration and 1.5 μL GlycoBlue to this sample (eluate), the “input” sample (collected in **step 1** in Subheading 3.4) and “supernatant” sample (collected in **step 5** in Subheading 3.4) and mix well. Precipitate RNA by adding 2.5 volumes of ice-cold 95% ethanol and incubating overnight at −20 °C.
2. Pellet RNA at 16,100 × *g* for 10 min at room temperature. Wash pellet with 500 μL ice-cold 75% ethanol by vortexing for 10 s. Centrifuge at 16,100 × *g* at room temperature for an additional 10 min. Remove ethanol by decanting, centrifuge again for 10 s at 16,100 × *g* at room temperature to collect any residual ethanol, and remove with micropipette. Air dry pellet until residual ethanol has evaporated.
3. Resuspend RNA pellet in 450 μL LET Buffer + 50 μL 10% SDS (*see* **Note 27**) by vortexing for 5 min at room temperature.
4. Reverse chemical cross-links by heating the sample at 65 °C for 1 h (*see* **Note 28**).

5. Extract RNA with phenol/chloroform by adding an equal volume (i.e., 500 μL) of P/C/L to each sample. Vortex 5 min at room temperature and separate phases by centrifugation at 16,100 × *g* for 5 min at room temperature. Carefully draw off the aqueous (top) layer and transfer it to a new 1.5 mL Eppendorf tube. Repeat extraction, substituting the P/C/L with an equal volume of chloroform.
6. Precipitate RNA overnight as in **steps 1** and **2** in Subheading 3.6.
7. Resuspend RNA in 20 μL LET Buffer for northern blot analysis, or 20 μL RNase-free dH_2O for qRT-PCR analysis (*see* **Note 29**).

3.7 Protein Analysis by SDS-PAGE (Optional) (See Notes 16, 23, and 25)

1. Concentrate proteins in eluate from **step 3** in Subheading 3.5 by evaporation under vacuum (using a SpeedVac or similar) on high heat until volume is reduced to ~20 μL (*see* **Note 30**). If multiple samples have been processed in parallel to increase scale, they may be combined prior to concentration at this step.
2. Precipitate proteins from concentrated eluate (and "input" and/or "supernatant" samples from **steps 1** and **5**, Subheading 3.4, if collected), by adding 4 volumes of ice-cold 100% acetone and incubating for 1 h at −20 °C.
3. Collect precipitated proteins by centrifugation for 10 min at 16,100 × *g* at 4 °C. Remove the supernatant, taking care to not disrupt the pellet.
4. Wash the protein pellet with ice-cold 80% acetone by vortexing for 60 s.
5. Repeat centrifugation as in **step 3** in Subheading 3.7. Remove the supernatant and air dry pellet to complete dryness (*see* **Note 31**).
6. Resuspend pellet in volume of 1× SDS Sample Buffer suitable for loading the full quantity into a single well of a SDS polyacrylamide gel (e.g., 20 μL) by vortexing for 5 min or until pellets are completely resuspended (*see* **Note 32**).
7. Heat proteins at 70 °C for 1 h to reverse chemical cross-links (*see* **Note 28**) followed by 95 °C for 5 min. Sample can now be analyzed by SDS-PAGE.

3.8 Protein Analysis by Mass Spectrometry

1. Resuspend beads in up to 100 μL Wash Buffer 2. Remove an aliquot of resuspended beads for RNA recovery analysis (ideally 1/10 volume) and process this sample (i.e., "eluate") with the "input" and supernatant' samples beginning at **step 2** in Subheading 3.5 (*see* **Note 16**). The remainder of the on-bead sample resuspended in Wash Buffer 2 can be stored at −80 °C.

2. Purified mRNP sample can be further processed for downstream analysis by mass spectrometry (i.e., on-bead trypsin digestion and mass spectrophotometric analysis).

4 Notes

1. Media, growth conditions, and lysis method are specific for the budding yeast, *S. cerevisiae*. Cell-type-specific methods could be substituted, in principle, but should be tested for compatibility.
2. In the established system, cells expressed a heterologous *GFP* reporter from a high-copy plasmid also harboring *URA3* (complementing the uracil auxotrophy of the yeast strain). Thus, uracil was omitted from the Amino Acid Dropout Mix.
3. The optimized protocol was performed using a yeast strain lacking three vacuolar proteases (*MAT*α, *ade2-1*, *ura3*, *leu2*, *his3*, *pep4Δ*::*HIS3*, *prb1*Δ::*his3*, *prc1*Δ::*hisG*) to reduce protein degradation during the procedure (*see* **Note 20**). Target mRNA can be expressed from its endogenous chromosomal locus, or, alternatively, from a high-copy plasmid (2 μ origin of replication and harboring a gene marker for plasmid maintenance) to enhance mRNA expression and mRNP yield. In order to identify protein components specifically bound to the target mRNA, mRNP purification should also be performed on cells lacking the target transcript, such as those with the endogenous gene locus deleted or lacking the plasmid-encoded copy of the target gene. Alternatively, for essential genes, the use of a repressible promoter could be used to inhibit target expression. We have found that performing the mRNP purification protocol using cells that do not express target mRNA (i.e., no plasmid control) yielded negligible protein contamination as determined by mass spectrometry.
4. It is critical to use RNase-free dH_2O for all reagents throughout the purification protocol beginning at cell lysis.
5. Additional magnetic Dynabeads options were tested (Fig. 2a). The MyOne Streptavidin C1 Dynabeads demonstrated the highest binding capacity for this purification scheme (Fig. 2b).
6. DNA oligonucleotides complementary to the target transcript should be individually tested for their ability to hybridize to and capture the RNA target. The recommended design for an oligonucleotide includes ~25 nucleotides of complementarity to the RNA of interest with a melting temperature of ~55 °C, a 15 nucleotide linker with no significant complementarity to genomic regions in the organism of interest, and a 3′ biotinylation modification. Oligonucleotides should be evaluated in

a

Magnetic bead type	Product #	Diameter	Surface character
M-270	65305	large	hydrophilic
M-280	11205D	large	hydrophobic
MyOne Streptavidin C1	65001	small	hydrophilic
MyOne Streptavidin T1	65601	small	hydrophobic

b

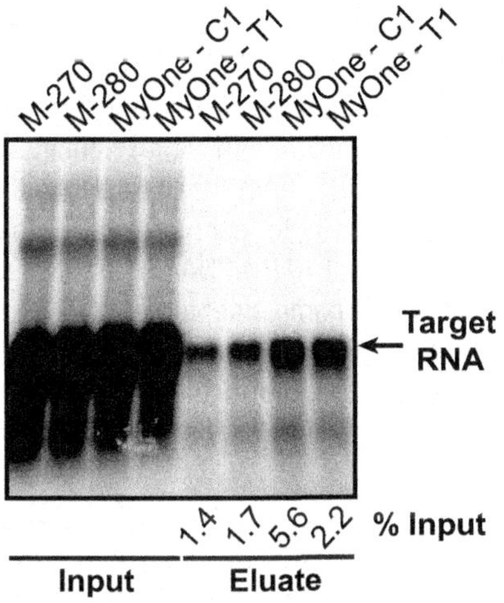

Fig. 2 Screening streptavidin-conjugated magnetic beads for RNA capture efficiency. (**a**) Table highlighting differences between magnetic bead systems tested. All beads are streptavidin-conjugated Dynabeads (ThermoFisher Scientific; product numbers indicated). Bead diameter: large = 2.8 μm; small = 1.05 μm. (**b**) Northern blot analysis of *GFP* mRNA (target RNA) captured from total purified yeast RNA using a single complementary DNA oligonucleotide conjugated to each magnetic bead system. Percent recovery of RNA from input is indicated below

silico to ensure that no substantial intramolecular secondary structure can form. We have observed that, similar to PCR primers or probes, not all complementary oligonucleotides are equally effective, even when the above guidelines are followed. Thus, it is highly recommended that each oligonucleotide be tested individually for the ability to purify the RNA species of interest. The combination of several oligonucleotides with distinct regions of complementarity to the target is likely to be the most effective at depleting the RNA from the lysate (Fig. 3a, b).

7. Culture size can be adjusted; however, the downstream lysis procedure is optimized for a cell pellet resulting from a 200 mL

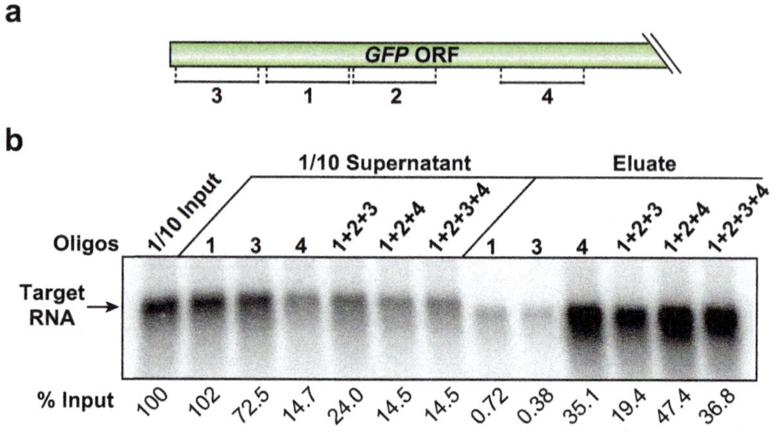

Fig. 3 Screening of complementary DNA oligonucleotides for efficiency of affinity capture of *GFP* mRNA. (**a**) Schematic of the 5′ region of the *GFP* open reading frame (ORF; from the start codon through codon 69). Approximate site of complementarity for each tested DNA oligonucleotide is indicated. (**b**) Northern blot analysis of *GFP* mRNA (target RNA) captured from total purified yeast RNA using different complementary DNA oligonucleotides (oligos) either alone or in combination. Oligonucleotide numbers correspond to the labels in (**a**). Percent target RNA remaining in the supernatant or captured in the eluate are quantified relative to the input and are indicated below

culture grown to an OD_{600} = 0.4. If increasing the scale of the experiment, it is recommended to harvest multiple cell pellets equivalent to a 200 mL culture and process in parallel. 200 mL of yeast culture is sufficient for downstream RNA analysis by northern blot or protein analysis by western blot; approximately 12 L of total cell volume was necessary and sufficient for three to four technical replicates of analysis by mass spectrometry.

8. Depending on the particular mRNP complex of interest, formaldehyde cross-linking conditions may need to be optimized. This may include altering the concentration and/or duration of treatment. Possible ways to monitor the extent of cross-linking include measuring overall protein yield from cell extracts (yield decreases as cross-linking increases), or testing for the loss of protein epitopes by western blotting (increased cross-linking often results in the masking of exposed epitopes).
9. Ensure that all cell lysis steps are carried out on ice and with reagents prechilled to 4 °C.
10. Resuspension of cell pellets stored at −80 °C may be difficult. It is recommended that samples be vortexed in short (~30 s) bursts and then returned to ice to ensure they remain at 4 °C. Proceed to the next step once no solid pellet is visible and the cell suspension appears uniform.

11. Passage of lysates through the opening created by puncturing the tube with an 18-gauge needle allows physical separation of the lysate from the glass beads and increases yield of the supernatant. Low-speed centrifugation of lysates ensures minimal loss of cross-linked protein complexes while removing cellular debris.
12. A second sequential lysis of the cellular material was found to significantly increase the yield and concentration of the cell extract. This is particularly important for formaldehyde cross-linked cells, which typically give rise to a lysate of lower concentration.
13. Quantification of cell extracts at this step is useful for monitoring the reproducibility of lysis from one sample to another. However, normalization of recovered material between samples or replicates is not performed using values acquired at this step.
14. Wash buffers and procedures follow the manufacturer's protocol for Dynabeads (ThermoFisher Scientific). While the manufacturer's protocol also recommends a set of bead washes with Solution A (0.1 M NaOH, 0.05 M NaCl) and Solution B (0.1 M NaCl) for applications involving RNA purification, our experience demonstrated no RNase contamination when these steps were eliminated. In contrast, we observed an increase in sample degradation when these steps were included, perhaps due to unintentional carryover of NaOH from Solution A (data not shown). Therefore, we recommend bypassing these wash steps (which have been omitted here).
15. If using multiple DNA oligonucleotides in combination, the total volume added at this step should remain at 25 μL and the combined concentration of oligonucleotides should total 10 nmol. Exceeding this total concentration does not result in increased target mRNA capture.
16. It is recommended that during both initial optimization of this protocol and for quality control during the experiment, aliquots of sample from each step of the procedure (input, supernatant, and eluate) be monitored for RNA or protein recovery, or both. Additionally, this step allows for the calculation of percent recovery and a determination of the integrity of the sample. Detailed procedures for sample processing to carry out RNA or protein analysis are described below.
17. The same reaction can be scaled to 10 mL in a single tube (i.e., extracts from two 200 mL cell pellets combined, also doubling the volume of Hybridization Reaction and beads) with no substantial loss of the capture of target transcript. Increasing the scale further was seen to result in a substantial decrease in the efficiency of RNA recovery. To increase the scale beyond

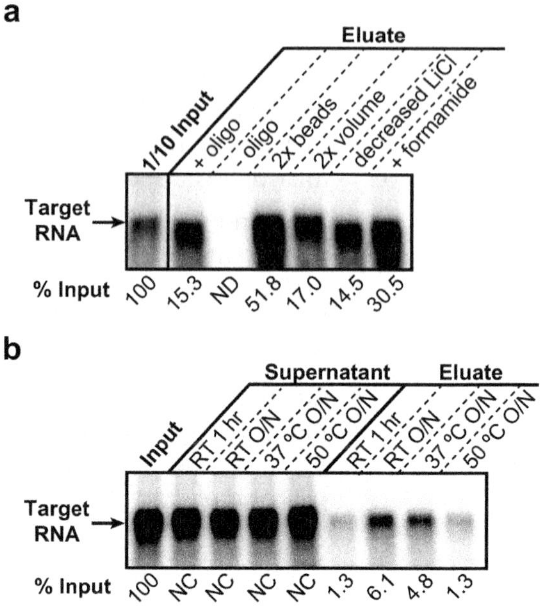

Fig. 4 Optimization of conditions for transcript-specific affinity capture. (**a**) Northern blot analysis of *GFP* mRNA (target RNA) captured from total purified RNA under varying hybridization conditions. Variables shown include bead volume, hybridization volume, salt concentration, and the addition of formamide to hybridization reactions. Percent recovered in eluate relative to input indicated below. *ND* not detected. Note, a number of additional optimization conditions were tested (not shown) and all conditions leading to increased capture of the target RNA were incorporated into the final protocol. (**b**) Northern blot analysis of *GFP* mRNA (target RNA) captured from total purified RNA under varying hybridization times and temperatures. *RT* room temperature, *O/N* overnight, *NC* not calculated

10 mL, it is recommended to carry out multiple purifications in parallel and combine samples prior to the final downstream analysis step selected.

18. Hybridization conditions were optimized to maximize mRNP purification while minimizing nonspecific binding (Fig. 4a). This is accomplished through the use of a high salt concentration (500 mM LiCl) to decrease nonspecific protein binding but promote DNA:RNA hybridization, ionic detergent (0.5% SDS) to decrease nonspecific protein binding (with no expected effect on hybridization), and 14% formamide to increase the stringency of DNA:RNA hybridization. EDTA is included to chelate divalent cations, reducing the possibility of RNA degradation during the incubation.

19. Overnight incubation at room temperature was found to result in the highest recovery of RNA from cell extracts (Fig. 4b). EDTA and SDS were sufficient to inhibit degradation of RNA in samples under these conditions. For mRNP purification from other cell types, the addition of a general RNase inhibitor may be warranted.
20. Protein degradation may occur in samples incubated overnight at room temperature. A combination of a protease-deficient yeast strain and a fungal-specific protease inhibitor was sufficient to inhibit protein degradation in our samples. It is recommended that an organism-specific protease inhibitor be used for samples not originating from yeast cells.
21. Wash Buffer 1 removes excess salt and SDS from the sample, both of which can have adverse effects on downstream applications such as mass spectrometry. Additionally, decreasing the salt concentration during the wash steps increases the stringency of DNA:RNA hybridization, helping to eliminate mRNP complexes that may result from off-target hybridization.
22. Wash Buffer 2 removes additional excess salt and SDS, and further increases DNA:RNA hybrid stringency (*see* **Note 21**). It is particularly important to ensure that the supernatant from each of these washes is removed completely in order to minimize residual SDS in the sample.
23. Steps in Subheadings 3.5–3.7 are recommended to monitor the efficiency of target RNA capture and the integrity of the purified RNA samples. These steps are performed on the "input" and "supernatant" aliquots and the resulting purified mRNP from **step 8** in Subheading 3.4.
24. Elution temperature was optimized to maximize recovery (based on the melting temperature of the DNA:RNA hybrid) without degradation of the RNA. This temperature may need to be adjusted if oligonucleotides have a different melting temperature than what is recommended here.
25. Protein analysis of a fraction of the purified eluate is not recommended, as the quantity is typically low and difficult to detect by western blotting. To analyze individual purified protein components of the captured mRNP complex, it is necessary to use the entire sample (and proceed to steps in Subheading 3.7).
26. Alternative RNA isolation methods may be substituted. It is recommended, however, to begin any RNA isolation protocol with an ethanol precipitation step in order to remove residual Hybridization Reaction or Wash Buffer components that may interfere with downstream isolation or analysis.

27. The addition of SDS serves to improve resuspension of the RNA/protein pellet, which still contains cross-linked material.
28. A lower temperature is recommended for reversing chemical cross-links when RNA analysis is being performed. The high heat used to reverse cross-links in samples for downstream protein analysis (**step 7** in Subheading 3.7) is likely to result in RNA degradation.
29. Both northern and qRT-PCR analyses of input, supernatant, and eluate aliquots are appropriate for determining percent yield and enrichment during purification. Northern analysis is recommended to monitor the integrity of the RNA samples.
30. Concentration of the protein sample is critical at this step as this protocol results in an eluate with low protein levels that, without concentration, are typically immeasurable by standard biochemical techniques. Tests using acetone to precipitate the entire unconcentrated sample (with or without carriers) indicated that substantial protein loss occurs. Loss is minimized by acetone precipitation of concentrated samples. Depending on the volume of sample being concentrated, reducing the volume to ~20 μL may take several hours.
31. Take care to ensure that the protein pellet does not over-dry, as this makes the protein much more difficult to resuspend.
32. Resuspension of protein pellet by vortexing may require longer than the suggested 5 min, and can be aided by micropipette mixing and/or gentle heat (37 °C). During resuspension, it is also recommended to run buffer along sides of tube to ensure entire sample is fully resuspended and material adhering to sides of tube is not lost.

Acknowledgments

We thank Tim Nilsen and members of his lab for providing helpful suggestions during the development of this method. Jeff Coller and Coller lab member, Najwa Alhusaini, provided helpful comments and critical reading of this manuscript. Mass spectrometry and identification of protein components within our transcript-specific mRNPs was achieved in collaboration with Amber Mosley and Whitney Smith-Kinnaman in the Department of Biochemistry and Molecular Biology at the Indiana University School of Medicine. This work was supported by funding by the National Institute of General Medical Sciences (GM095621 to K.E.B.; T32 GM008056 to J.E.S.) and the National Science Foundation (NSF1253788 to K.E.B.).

References

1. Muller-McNicoll M, Neugebauer KM (2013) How cells get the message: dynamic assembly and function of mRNA-protein complexes. Nat Rev Genet 14(4):275–287. doi:10.1038/nrg3434
2. Singh G, Pratt G, Yeo GW, Moore MJ (2015) The clothes make the mRNA: past and present trends in mRNP fashion. Annu Rev Biochem 84:325–354. doi:10.1146/annurev-biochem-080111-092106
3. McHugh CA, Russell P, Guttman M (2014) Methods for comprehensive experimental identification of RNA-protein interactions. Genome Biol 15:203. doi:10.1186/gb4152
4. Riley KJ, Steitz JA (2013) The "Observer Effect" in genome-wide surveys of protein-RNA interactions. Mol Cell 49(4):601–604. doi:10.1016/j.molcel.2013.01.030
5. Mili S, Steitz JA (2004) Evidence for reassociation of RNA-binding proteins after cell lysis: implications for the interpretation of immunoprecipitation analyses. RNA 10(11):1692–1694. doi:10.1261/rna.7151404
6. Riley KJ, Yario TA, Steitz JA (2012) Association of Argonaute proteins and microRNAs can occur after cell lysis. RNA 18(9):1581–1585. doi:10.1261/rna.034934.112
7. Darnell R (2012) CLIP (cross-linking and immunoprecipitation) identification of RNAs bound by a specific protein. Cold Spring Harb Protoc 2012(11):1146–1160. doi:10.1101/pdb.prot072132
8. Van Nostrand EL, Huelga SC, Yeo GW (2016) Experimental and computational considerations in the study of RNA-binding protein-RNA interactions. Adv Exp Med Biol907:1–28.doi:10.1007/978-3-319-29073-7_1
9. Bardwell VJ, Wickens M (1990) Purification of RNA and RNA-protein complexes by an R17 coat protein affinity method. Nucleic Acids Res 18(22):6587–6594
10. Hogg JR, Collins K (2007) RNA-based affinity purification reveals 7SK RNPs with distinct composition and regulation. RNA 13(6):868–880. doi:10.1261/rna.565207
11. Walker SC, Scott FH, Srisawat C, Engelke DR (2008) RNA affinity tags for the rapid purification and investigation of RNAs and RNA-protein complexes. Methods Mol Biol 488:23–40. doi:10.1007/978-1-60327-475-3_3
12. Said N, Rieder R, Hurwitz R, Deckert J, Urlaub H, Vogel J (2009) In vivo expression and purification of aptamer-tagged small RNA regulators. Nucleic Acids Res 37(20):e133. doi:10.1093/nar/gkp719
13. Yoon JH, Gorospe M (2016) Identification of mRNA-interacting factors by MS2-TRAP (MS2-tagged RNA affinity purification). Methods Mol Biol 1421:15–22. doi:10.1007/978-1-4939-3591-8_2
14. Baltz AG, Munschauer M, Schwanhausser B, Vasile A, Murakawa Y, Schueler M, Youngs N, Penfold-Brown D, Drew K, Milek M, Wyler E, Bonneau R, Selbach M, Dieterich C, Landthaler M (2012) The mRNA-bound proteome and its global occupancy profile on protein-coding transcripts. Mol Cell 46(5):674–690. doi:10.1016/j.molcel.2012.05.021
15. Castello A, Fischer B, Eichelbaum K, Horos R, Beckmann BM, Strein C, Davey NE, Humphreys DT, Preiss T, Steinmetz LM, Krijgsveld J, Hentze MW (2012) Insights into RNA biology from an atlas of mammalian mRNA-binding proteins. Cell 149(6):1393–1406. doi:10.1016/j.cell.2012.04.031
16. Mitchell SF, Jain S, She M, Parker R (2013) Global analysis of yeast mRNPs. Nat Struct Mol Biol 20(1):127–133. doi:10.1038/nsmb.2468
17. Matia-Gonzalez A, Laing E, Gerber A (2015) Conserved mRNA-binding proteomes in eukaryotic cells. Nat Struct Mol Biol 22(12):1027–1033. doi:10.1038/nsmb.3128
18. Beckmann BM, Horos R, Fischer B, Castello A, Eichelbaum K, Alleaume AM, Schwarzl T, Curk T, Foehr S, Huber W, Krijgsveld J, Hentze MW (2015) The RNA-binding proteomes from yeast to man harbour conserved enigmRBPs. Nat Commun 6:10127. doi:10.1038/ncomms10127
19. Castello A, Fischer B, Frese CK, Horos R, Alleaume AM, Foehr S, Curk T, Krijgsvelt J, Hentze MW (2016) Comprehensive identification of RNA-binding domains in human cells. Mol Cell 63(4):696–710. doi:10.1016/j.molcel.2016.06.029
20. Liao Y, Castello A, Fischer B, Leicht S, Foehr S, Frese CK, Ragan C, Kurscheid S, Pagler E, Yang H, Krijgsvelt J, Hentze MW, Preiss T (2016) The cardiomyocyte RNA-binding proteome: links to intermediary metabolism and heart disease. Cell Rep 15(5):1456–1469. doi:10.1016/j.celrep.2016.06.084
21. Chu C, Zhang QC, da Rocha ST, Flynn RA, Baradwaj M, Calabrese JM, Magnuson T, Heard E, Chang HY (2015) Systematic discovery of Xist RNA binding proteins. Cell 161(2):404–416. doi:10.1016/j.cell.2015.03.025

22. McHugh CA, Chen CK, Chow A, Surka CF, Tran C, McDonel P, Pandya-Jones A, Blanco M, Burghard C, Moardian A, Sweredoski MJ, Shishkin AA, Su J, Lander ES, Hess S, Plath K, Guttman M (2015) The Xist lncRNA interacts directly with SHARP to silence transcription through HDAC3. Nature 521(7551):232–236. doi:10.1038/nature14443
23. Niranjanakumari S, Lasda E, Brazas R, Garcia-Blanco MA (2002) Reversible cross-linking combined with immunoprecipitation to study RNA-protein interactions in vivo. Methods 26(2):182–190. doi:10.1016/S1046-2023(02)00021-X
24. Smith JE, Whiteside DL, Smith-Kinnaman WR, Mosley AL, Baker KE (In preparation) mRNP analysis of an NMDsensitive mRNA in yeast implicates UPF1 as the sensor for substrate recognition

Chapter 16

A Cell-Based High-Throughput Method for Identifying Modulators of Alternative Splicing

Sika Zheng

Abstract

Alternative splicing, a key regulatory process of gene expression, is controlled by *trans*-acting factors that recognize *cis*-elements in premature RNA transcripts to affect spliceosome assembly and splice site choices. Extracellular stimuli and signaling cascades can converge on RNA binding splicing regulators to affect alternative splicing. Defects in splicing regulation have been associated with various human diseases, and modification of disease-causing splicing events presents great therapeutic promise. Determining splicing regulators and/or upstream modulators has been difficult with low throughput, low sensitivity, and low specificity. IRAS (identifying regulators of alternative splicing) is a novel cell-based high-throughput screening strategy designed specifically to address these challenges and has achieved high throughput, high sensitivity, and high specificity. Here, we describe the IRAS method in detail with a pair of dual-fluorescence minigene reporters that produces GFP and RFP fluorescent signals to assay the two spliced isoforms exclusively. These two complementary mini-gene reporters alter GFP/RFP output ratios in the opposite direction in response to only a true splicing change. False positives from a signal screen do not stimulate opposite changes in GFP/RFP ratios. The reporter pair in conjunction with robotic liquid handlers and arrayed libraries allows IRAS to screen for both positive and negative splicing regulators with high sensitivity and specificity in a high-throughput manner.

Key words Dual-fluorescence, Dual-output, Minigene, Splicing reporters, High-throughput screen, Alternative splicing, GFP, RFP, Splicing regulator, IRAS

1 Introduction

Many pre-mRNA alternative splicing events are highly controlled to produce functionally distinct gene products during development or in response to extracellular stimuli [1–3]. A single alternative exon typically contains multiple *cis*-elements in itself or its surrounding introns, and is thus regulated by multiple cognate *trans*-factors [4, 5]. On the other hand, each *trans*-factor can affect many target exons [6]. Despite extensive studies on a dozen of RNA binding proteins that have been characterized as alternative splicing regulators in eukaryotic cells, many more await discovery

Yongsheng Shi (ed.), *mRNA Processing: Methods and Protocols*, Methods in Molecular Biology, vol. 1648,
DOI 10.1007/978-1-4939-7204-3_16,

to fill in the large alternative splicing regulatory network that impact almost every multi-exon gene in all aspects of biology. Factors modulating the activity of RNA binding splicing regulators are also poorly understood.

Global identification of exon targets for individual splicing factor has been aided by genome-wide sequencing-based methods [7–10]. However, identification of splicing regulators for a specific alternative exon is generally difficult and intensively laborious. Successful screens were reported isolating single splicing factor, but were mostly insensitive for broader discovery of regulatory factors [11–13]. Most of these methods used a single-output splicing reporter for which splicing changes were measured by changes in expression of one of the two isoforms. Transcriptional activity of the reporters apparently contributed to the readout of these single-output reporters, and thus led to a multitude of false positives. Additionally, the single-output reporters had to be spliced at very low basal levels to obtain a large signal-to-noise ratio for the screens. The use of a pair of single-output reporters could overcome some of these limitations [14], but was confronted with new variables resulting from unpredictable integration of two minigene reporters into genomes.

False positive hits modulating transcriptional activity in screening can be mitigated by dual-output reporters in which both splicing isoforms are assayed allowing for screening of changes in isoform ratio. Furthermore, the expression of the two alternative isoforms is negatively correlated, so using the ratio of the two isoforms to measure splicing level squares the dynamic range given by individual output. Various versions of dual-fluorescence splicing reporters have been successfully used to identify modulators of splicing [15, 16]. However, variables affecting the readout (i.e., the output ratio of the two isoforms) without an impact on pre-mRNA splicing still constituted a majority of (often about 90%) hits, probably because splicing regulation in a form of splicing ratio was typically moderate in comparison to many other gene regulation steps.

Our strategy, IRAS, uses two complementary dual-output minigene reporters that exhibit opposite changes of the readouts in response to splicing alternation (Fig. 1a). Either fluorescence-based or luminescence-based readout is compatible with IRAS. We herein describe the IRAS method using proven fluorescence-based reporters. The first reporter (pflareG) produces GFP when an exon is included and RFP when the exon is skipped, while a second reporter (pflareA) produces RFP and GFP to represent exon inclusion and exclusion respectively (Fig. 1a). Therefore, an increase in exon inclusion is reflected by an increase in the GFP/RFP ratio in the plareG reporter but a decrease in the GFP/RFP ratio in the pflareA vector. The IRAS strategy screens the same library with the two reporters in parallel. A true splicing modulator should

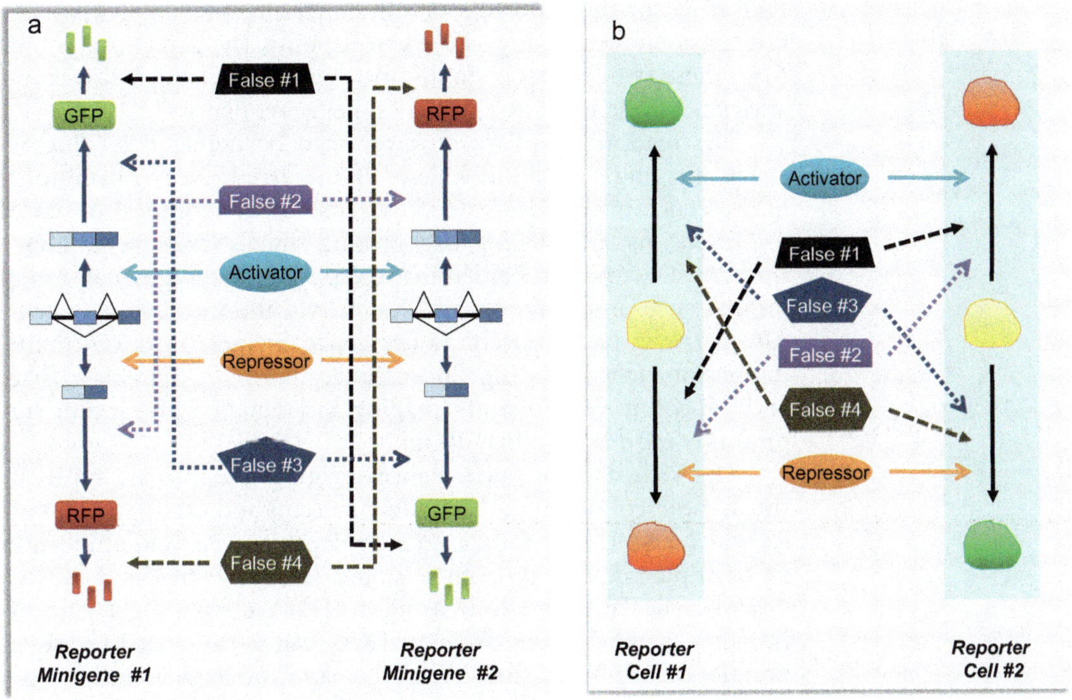

Fig. 1 Illustration of the IRAS strategy using two dual-fluorescence minigene reporters. (**a**) Two complementary splicing reporters (#1 and #2) differ in only a few nucleotides. Reporter #1 translates GFP when the alternative exon is included and RFP when the alternative exon is skipped. Reporter #2 translates RFP from the included isoform and GFP from the skipped isoform. The inclusion ratios of the alternative exon in reporter #1 and #2 are therefore expressed by the ratios of GFP/RFP and RFP/GFP respectively. Factors altering these ratios without affecting splicing are false positives in the screens. False positives could differentially promote translation efficiency of GFP (e.g., #3) or RFP (e.g., #2), or alternatively, promote degradation of GFP (e.g., #1) or RFP (e.g., #4). (**b**) True splicing activators and repressors or false positive hits from screening induce differential phenotypes of shifting the GFP/RFP ratios in the two reporter cells. A true splicing regulator changes GFP/RFP ratios in the two reporter cells in the opposite direction, whereas a false positive more likely affects the GFP/RFP ratios in the same direction

change the GFP/RFP ratios of the two reporters in the opposite direction (Fig. 1b). For example, a splicing activator increases the GFP/RFP ratio in the pflareG reporter but decreases the GFP/RFP ratio in the pflareA reporter. A typical false positive from a single screen either preferentially stabilizes one of the fluorescent proteins (False #1 or #4 in Fig. 1a) or differentially affects ORF translation efficiency between GFP and RFP (False #2 or #3 in Fig. 1a). This false positive changing the GFP/RFP ratio of the first minigene is very unlikely to change the GFP/RFP ratio of the second minigene in the opposite direction, because pflareA and pflareG minigenes differ in only a few nucleotides. Instead, this false positive more likely changes GFP/RFP ratios of the two reporters in the same direction when affecting only either GFP or

RFP ORF of the reporters or only affecting the stability of one fluorescent protein (Fig. 1b). Overlapping the screening results from these two reporters significantly eliminates false positives and enriches for identification of true splicing regulators.

The readout of pflare reporters can be detected by fluorescence imaging. Typhoon imager is preferred over conventional fluorescence microplate readers because of the confocal optics of Typhoon imager reducing background signals from media above the cells. Data points yielded from Typhoon imaging equal overall fluorescence intensity of all the cells within individual wells. Alternatively, high-content fluorescence imager produces multi-dimension data including fluorescence intensity, cell shape, size, and cell count. However, its imaging and data processing time significantly reduce the throughout.

The versatility of fluorescence imaging is enhanced by high-throughput robotic liquid-handling systems and catalogued libraries which enable screening at greater depth as well as identification of multiple regulators from a single screen. Here in detail we describe the IRAS method using a cDNA library for a gain-of-function genetic screen (Fig. 2). IRAS can also be used to screen libraries of siRNAs or shRNAs, and of small molecules, and is easily adaptable for any cassette exon of interest for genome-wide identification of its regulatory factors [17].

2 Materials

2.1 Construction of Dual-Fluorescence Minigene Reporters

- pflareA backbone reporter.
- plfareG backbone reporter.
- Phusion High-Fidelity DNA polymerase.
- Purified genomic DNA containing the gene and exon of interest.
- QuikChange II XL Site-Directed Mutagenesis Kit or similar.
- QIAquick Gel Extraction Kit or similar.
- NanoDrop.
- Alkaline phosphatase.
- T4 DNA Ligase.
- TOP10 Competent Cells.
- EcoRI.
- BamHI.
- LB-kanamycin agar plate.
- Plasmid Miniprep Kit.
- Agarose gel electrophoresis setup.

2.2 Generation of Stable Cell Clones

- A parental cell line of choice.
- QIAquick Gel Extraction Kit.

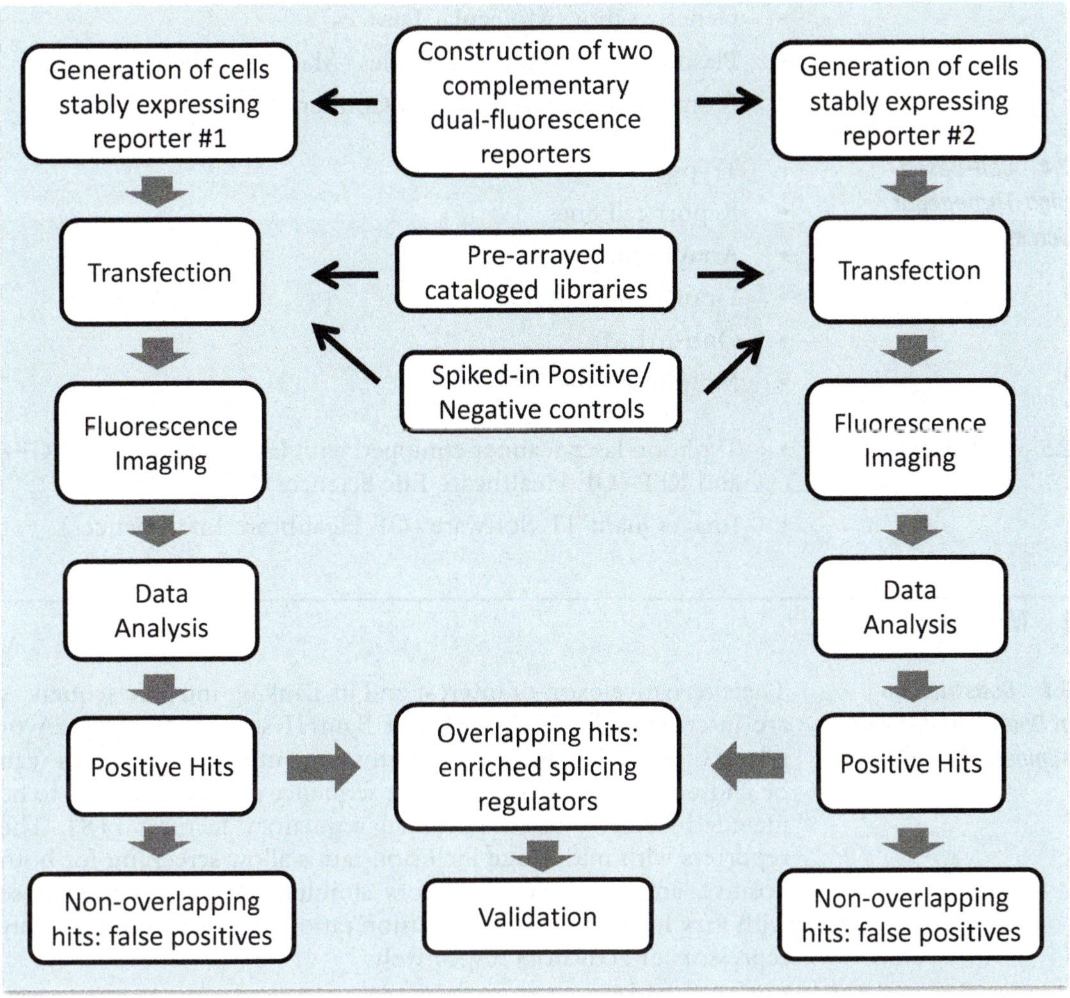

Fig. 2 Workflow of a fluorescent cell-based high-throughput genetic screen using the IRAS strategy

- DraIII.
- Lipofectamine 2000.
- G418 or Geneticin.
- Phenol red free media.
- FACS machine.

2.3 Library and Array Construction

- The Mammalian Gene Collection (MGC) library: developed by NIH and representing the most extensive collection of mammalian cDNA clones available. All cDNA clones are in the pCMVsport6.0 or pCMVsport6.1 backbone vectors pre-arrayed in 96-well plates (Life Technology).
- 384-well glass-bottom plates.

- Genetix Qbot (Molecular Devices).
- Plasmid mini-prep consumables (Macherey-Nagel).
- Biomek FX robot (Beckman Coulter).

2.4 Cell-Based High-Throughput Screens

- Trypsin.
- Report cell lines.
- Arrayed library plates.
- Lipofetamine 2000.
- Opti-MEM.
- Multidrop 384.

2.5 Data Acquisition

- Typhoon laser scanner equipped with lasers and filters for GFP and RFP (GE Healthcare Life Sciences).
- ImageQuant TL Software (GE Healthcare Life Sciences).

3 Methods

3.1 Construction of Dual-Fluorescence Minigene Reporters

The alternative exon of interest and its flanking intronic sequences are inserted into the EcoRI and BamHI sites of the pflareA or pflareG vectors. Different lengths of intronic sequences can be chosen, but conserved intronic sequence is recommended to be included because of its potential regulatory features [18]. The reporters with mid-range inclusion ratios allow screening for both positive and negative regulators simultaneously, whereas those with very high or very low inclusion ratios allow screening for only repressors or activators respectively.

1. Design forward and reverse primers for cloning the alternative exon of interest and its flanking introns.
2. Synthesize the forward primer with an EcoRI recognition site at the 5′ end. Synthesize the reverse primer with a BamHI recognition site at the 5′ end.
3. PCR amplification using the above forward and reverse primers, Phusion high-fidelity DNA polymerase, and genomic DNA as the template.
4. Confirm the PCR amplicon by agarose gel electrophoresis.
5. Excise the band of the PCR amplicon and purify the PCR product with QIAquick Gel Extraction Kit.
6. Measure the concentrations of the recovered PCR products with NanoDrop.
7. Digest the PCR product with EcoRI and BamHI simultaneously for 1 h at 37 °C. Meanwhile linearize the pflareA and pflareG vectors with EcroRI and BamHI for 1 h at 37 °C.

8. Dephosphorylate the linearized vectors with alkaline phosphatase for 0.5 h at 37 °C.
9. Confirm the linearized backbone vectors by agarose gel electrophoresis.
10. Gel-purify the digested PCR products and vectors using QIAquick Gel Extraction Kit.
11. Ligate the PCR product to the linearized vector with T4 DNA ligase at 16 °C overnight.
12. Transform the ligated products into TOP10 competent cells.
13. Plate the transformed competent cells on LB-kanamycin agar plates and incubate at 37 °C overnight.
14. Select several colonies for amplification and purification of pflareA-exon and plfareG-exon reporters using Plasmid Miniprep Kit.
15. For the plfareA-exon vector, mutate any ATG codons present within the alternative exon of interest to ATA using the QuikChange II XL Site-Directed Mutagenesis Kit.
16. For the pflareG-exon reporter, an ATG start codon within the alternative exon in frame with the GFP ORF is required for expression of GFP from exon-included transcripts. If missing, a start codon needs to be created using the QuikChange II XL Site-Directed Mutagenesis Kit. The Kozak consensus sequence gccRccATGG is also recommended to promote translation initiation for the GFP ORF.
17. Transform and purify the final versions of the pflareA-exon or pflareG-exon vectors. These vectors shall be tested for their reproducibility and applicability for the screen (*see* **Note 1**).

3.2 Generation of Stable Cell Clones

The choice of parental cells lines should meet the following criteria. The cell line ideally has characteristics of or is derived from the primary cells of interest. This is particularly important for loss-of-function screens because the most interesting cell/tissue type specific regulators sometimes express only in a highly relevant cell line. The cell line also needs to be highly transfectable for genetic screens to have reasonable signal-to-noise ratios.

1. Linearize the pflareA-exon vector by DraIII digestion and gel purify the linearized vector with QIAquick Gel Extraction Kit.
2. Transfect the linearized pflareA-exon vector into the target cell line of choice using Lipofectamine 2000.
3. Maintain and split the transfected cells in G418-containing media for 2 weeks to enrich stable cell clones expressing the reporter.
4. Sort double fluorescent (i.e., GFP + RFP+) single cells to individual wells of multi-well plates with a FACS machine and

allow cell clones to grow and expand in G418-containing media. Some cell clones may lose GFP and RFP expression during the selection process and can be discarded. From now on, the cells need to be maintained in phenol red-free media for enhanced fluorescence visualization and imaging.

5. Test multiple cell clones for their screening feasibility (*see* **Note 2**).
6. Repeat **steps 1–5** to generate stable cell clones expressing pflareG-exon reporter. The stable cell clones need to be confirmed for their responsiveness to splicing changes before high-throughput screens (*see* **Note 2**).

3.3 Library and Array Construction

1. Duplicate the MGC collection bacteria plates into 96-well plates using Genetix Qbot (Molecular Devices). This library contains about 16 thousand clones. Grow the culture overnight.
2. Prepare the plasmid DNA of the library using plasmid prep kit (Macherey-Nagel) with Biomek FX robot (Beckman Coulter).
3. Normalize plasmid concentration to 20 ng/ul for each well.
4. Spot the cDNA library into 384-well assay plates. We typically spot 40 ng cDNA (i.e., 2 μL) per 384-well (except the wells of A-H23 and A-H24).
5. Spot 40 ng negative control (pCMV sport 6.0) plasmids to wells of C23, C24, D23, and D24, 40 ng plasmids of a positive regulator (if available) to wells of E23, E24, F23, and F24, as well as a negative regulator (if available) to wells of G23, G24, H23, and H24 (16). If positive or negative control regulators are not available, leave the corresponding wells empty.
6. Two identical sets of "assay ready" library plates are prepared for pflareA and pflareG screens respectively. The assay plates take days to prepare and should be kept in −80 °C until use (*see* **Note 3**).

3.4 Cell-Based High-Throughput Screens

Cell number and transfection conditions shall be optimized (*see* **Note 4**). We have identified the optimal reverse-transfection condition for N2a cells using 8000 cells per 384-well with 40 ng plasmid DNA and 0.12 μL Lipofectamine 2000. One cDNA library screen encompassing 45384-well plates requires about 150 million cells and 2 mL Lipofetamine 2000.

1. Scale up the reporter cell cultures to ten 15 cm culture dishes by the day of performing the screens.
2. Thaw the cDNA plates while the reporter cells and transfection reagents are being prepared.
3. Dissociate the cells by trypsin.
4. Resuspend and dilute the cells in phenol-red-free feeding media to be 400 thousand cells per milliliter.

5. Mix 2 mL Lipofetamine 2000 with 166 mL Opti-MEM.
6. After the cDNA plates are thawed, dispense 10 μL Opti-MEM containing 0.12 μL Lipofectamine 2000 into each well with Multidrop 384 to mix with the plasmid DNA in the plates.
7. Dispense 8000 reporter cells in 20 μL feeding media to every well except A23, A24, B23, and B24 wells using Multidrop 384.
8. Dispense cell-free feeding media into A23, A24, B23, and B24 of all plates. These four wells will be used to derive a scale factor for inter-plate normalization (*see* below).
9. Dispense cell-free media to one (or multiple) new 384-well plate as a background plate to measure background GFP and RFP fluorescence intensity of the cell media.

3.5 Data Acquisition

1. 48 h after transfection, arrange and scan the plates on the Typhoon scanner using 200 μm pixel size as the scanning resolution for 384-well plates (*see* **Note 5**). GFP signals are acquired using the blue laser (488 nm) and 520 nm band pass emission filter. RFP signals are obtained using the green laser (532 nm) and 610 nm band pass emission filter. Keep the scanning parameters (including the PMT voltages) consistent across all plates.
2. Use ImageQuant TL software to analyze the scanned images (.gel files) and obtain raw GFP and RFP fluorescence intensity of every well.
3. Export the GFP and RFP signal values into Data/Text/Excel files.

3.6 Data Analysis

1. Wells A23, A24, B23, and B24 of each plate are used to derive a scaling factor for normalizing that plate. For example, XFP_{ijk} is the raw GFP or RFP intensities of a well at row "i" column "j" in plate k. Calculate the scale factor S_k for plate k based on the A23, A24, B23, and B24 wells by the following equation:

$$S_k = \frac{1}{4} \sum_{\substack{p=A,B \\ q=23,24}} \frac{\overline{XFP_{pq}}}{XFP_{pqk}}$$

where $\overline{XFP_{pq}}$ is the average values across all the plates.

2. The normalized and background-corrected GFP and RFP expression values (XFP_x) of the stable cells after transfection with cDNA x at row "i" column "j" in plate c are calculated with the following equation:

$$XFP_x = XFP_{ijc} S_c - \frac{1}{n_b} \sum_{b=1}^{n_b} XFP_{ijb} S_b$$

where S_c and S_b are the scale factors for the cell plates and the background plates respectively, and n_b is the total number of background plates used.

3. To estimate the splicing ratio of the reporter upon expression of cDNA x, use $\frac{GFP_x}{RFP_x}$ for the pflareG-exon minigene reporter and $\frac{RFP_x}{GFP_x}$ for the pflareA-exon minigene reporter.
4. To calculate the basal splicing level of the reporter, use the four wells (i.e., C23, C24, D23, and D24) that are transfected with empty control vector in every plate and derive the mean $\frac{GFP_{ctrl}}{RFP_{ctrl}}$ in the pflareG-exon cells and the mean $\frac{RFP_{ctrl}}{GFP_{ctrl}}$ in the pflareA-exon cells.
5. To examine the effect of cDNA x on the splicing of the pflareG reporter, use the following formula to approximate the change in the splicing ratio (M_x):

$$M_x = \log_2\left[\frac{GFP_x}{RFP_x}\right] - \log_2\left[\frac{GFP_{ctrl}}{RFP_{ctrl}}\right]$$

where $\frac{GFP_{ctrl}}{RFP_{ctrl}}$ is derived from the same plate as cDNA$_{x.}$ Similarly, to examine the action of cDNA x on the splicing of the pflareA reporter, calculate the change in the splicing ratio (M_x) as followed:

$$M_x = \log_2\left[\frac{RFP_x}{GFP_x}\right] - \log_2\left[\frac{RFP_{ctrl}}{GFP_{ctrl}}\right]$$

where $\frac{RFP_{ctrl}}{GFP_{ctrl}}$ is derived from the same plate as cDNAx. A value of $M_x > 0$ indicates a possible increase in splicing by cDNAx. A value of $M_x < 0$ indicates a possible decrease in splicing.
6. To assess the overall fluorescence expression of cells after transfection, or the "A" value, use the following equation:

$$A_x = \frac{1}{2}\left(\log_2\left(GFP_x\right) + \log_2\left(RFP_x\right)\right)$$

Multiple variables, including live cell numbers, transcriptional activity and translational activity of the reporter can have an impact on A values.
7. The M values, indicative of splicing changes, can be scatter plotted against the A values to visualize the performance of the whole library.

8. With the R package locfdr, calculate z score and local false discovery rate (FDR) for individual cDNA by comparing their M values to the population distribution of M values. We have found that a typical FDR threshold of 0.05 has yielded relatively high sensitivity (80–88%) for our reporter screens.
9. The two parallel screens using pflareA and pflareG reporters produce two different sets of possible splicing regulators. Overlap the two target sets to eliminate a large proportion of false positives. Proceed to validate the overlapping hits with secondary methods (*see* **Note 6**).

4 Notes

1. The two reporters, when expressing in cells, should exhibit similar splicing levels of the inserted alternative exons. They should also express both GFP and RFP, but with opposite preference. Any known modulator of the inserted alternative exon should change the GFP/RFP ratio of the reporters, and more importantly in the opposite direction for the two reporters. In addition to examining the performance of the reporters, these tests also confirm a known modulator suitable as a spike-in control in the final screen.
2. Reporter minigene expression varies substantially among stable cell clones. Cell clones of higher expression is not necessarily better, because they may not be as efficiently transfectable or responsive to known regulators. Furthermore, some clones can change GFP/RFP ratios even when transfected with an empty control vector as we have observed. Therefore, multiple clonal lines should be tested to find the optimal one.

   A good cell clone does not change GFP/RFP ratio when transfected with an empty control vector, and should do so in response to transfection with a known regulator of the exon. An RT-PCR assay to measure the splicing changes at the RNA level is also recommended. A good clone should also be efficiently transfectable. After transfection with a GFP expression plasmid, a highly transfectable cell clone shows very high GFP signals without affecting the RFP signals.
3. IRAS is broadly applicable to screen any library with considerations taken in light of the library attributes. Loss-of-function screens often require multiple shRNA or siRNA hits targeting the same gene to declare a true target. Because of functional compensation, regulatory paralogs may not be easily identified in loss-of-function screens. Loss-of-function screens may also miss true regulators that are absent in the screened cells. Gain-of-function screen has less concern of off-target effects, paralogous compensation, or missing expression, but overexpression can induce artificial splicing changes.

4. The screening parameters and the kinetics of reporter expression need to be examined for screening different libraries or in different cell types. For example, we have seen that cDNA overexpression plateaus at 2 days posttransfection in N2a cells but RNAi depletion typically requires 3 days or longer. To avoid overconfluency within a well, the number of cells initially plated needs to be optimized; this can affect the transfection condition.
5. To meet the similar data quality, pixel size of 100 μm should be used to screen 1536-well plates.
6. Overlapping candidates typically constitute about 5–20% of total hits. The candidates should be tested for their actions on the splicing of the endogenous exon, because minigene reporters do not necessarily recapitulate the regulation of endogenous exons. The minigene may not contain all of the relevant *cis*-regulatory elements. Molecules that affect the splicing of the minigene reporter need to be confirmed on endogenous transcripts in the relevant cell type. This can be achieved by either overexpressing the cDNA candidate or RNAi depletion of the candidate gene followed by RT-PCR assay of the endogenous spliced variants.

References

1. Chen M, Manley JL (2009) Mechanisms of alternative splicing regulation: insights from molecular and genomics approaches. Nat Rev Mol Cell Biol 10:741–754
2. Zheng S, Black DL (2013) Alternative pre-mRNA splicing in neurons: growing up and extending its reach. Trends Genet 29:442–448
3. Zheng S (2016) Alternative splicing and nonsense-mediated mRNA decay enforce neural specific gene expression. Int J Dev Neurosci. doi:10.1016/j.ijdevneu.2016.03.003
4. Black DL (2003) Mechanisms of alternative pre-messenger RNA splicing. Annu Rev Biochem 72:291–336
5. Fu X-D, Ares M Jr (2014) Context-dependent control of alternative splicing by RNA-binding proteins. Nat Rev Genet 15:689–701
6. Vuong CK, Black DL, Zheng S (2016) The neurogenetics of alternative splicing. Nat Rev Neurosci 17:265–281
7. Chen L, Zheng S (2009) Studying alternative splicing regulatory networks through partial correlation analysis. Genome Biol 10:R3
8. König J, Zarnack K, Rot G, Curk T, Kayikci M, Zupan B et al (2010) iCLIP reveals the function of hnRNP particles in splicing at individual nucleotide resolution. Nat Struct Mol Biol 17:909–915
9. Stork C, Zheng S (2016) Genome-wide profiling of RNA-protein interactions using CLIP-Seq. Methods Mol Biol 1421:137–151
10. Raj B, Blencowe BJ (2015) Alternative splicing in the mammalian nervous system: recent insights into mechanisms and functional roles. Neuron 87:14–27
11. Kar A, Havlioglu N, Tarn W-Y, Wu JY (2006) RBM4 interacts with an intronic element and stimulates tau exon 10 inclusion. J Biol Chem 281:24479–24488
12. Oberdoerffer S, Moita LF, Neems D, Freitas RP, Hacohen N, Rao A (2008) Regulation of CD45 alternative splicing by heterogeneous ribonucleoprotein, hnRNPLL. Science 321:686–691
13. Topp JD, Jackson J, Melton AA, Lynch KW (2008) A cell-based screen for splicing regulators identifies hnRNP LL as a distinct signal-induced repressor of CD45 variable exon 4. RNA 14:2038–2049
14. Kuroyanagi H, Ohno G, Sakane H, Maruoka H, Hagiwara M (2010) Visualization and genetic analysis of alternative splicing regulation in vivo using fluorescence reporters in transgenic *Caenorhabditis elegans*. Nat Protoc 5:1495–1517
15. Stoilov P, Lin C-H, Damoiseaux R, Nikolic J, Black DL (2008) A high-throughput screening

strategy identifies cardiotonic steroids as alternative splicing modulators. Proc Natl Acad Sci U S A 105:11218–11223

16. Kuroyanagi H, Kobayashi T, Mitani S, Hagiwara M (2006) Transgenic alternative-splicing reporters reveal tissue-specific expression profiles and regulation mechanisms in vivo. Nat Methods 3:909–915
17. Zheng S, Damoiseaux R, Chen L, Black DL (2013) A broadly applicable high-throughput screening strategy identifies new regulators of Dlg4 (Psd-95) alternative splicing. Genome Res 23:998–1007
18. Chen L, Zheng S (2008) Identify alternative splicing events based on position-specific evolutionary conservation. PLoS One 3:e2806

Chapter 17

Genome-Wide RNAi Screens for RNA Processing Events in *Drosophila melanogaster* S2 Cells

Natoya Peart and Eric J. Wagner

Abstract

Over the past 10 years, the design and application of genome-wide screening (GWS) has improved to the point that it can now be done at level of the individual laboratory. The advantages of GWSs compared to classical genetic screens include: immediate identification of a positive scoring gene, relatively short period of time necessary to conduct the screen (as little as 1 week), cell lines do not present developmental needs for gene expression that an organism normally would, and validation/confirmation of results is straightforward. Here, we describe a general protocol for GWS to be conducted in *Drosophila melanogaster* S2 cells. We provide specific details on what type of experiments must be done before initiating a screen, the materials that are required to conduct a screen, and make suggestions on methods to carry out secondary screening and counter-screening once the initial GWS is complete. Multiple considerations are also raised that focus on how to anticipate false positives/negatives and how to minimize their occurrence through intelligent design. Finally, we provide specific examples of data that our group has gathered from published genome-wide screens in order to exemplify how "hits" are scored and confirmed.

Key words High-throughput, RNAi, *Drosophila melanogaster*, Reporter-based, Gain-of-function, Loss-of-function

1 Introduction

RNA interference (RNAi) is a well-described process by which small double-stranded RNA molecules posttranscriptionally downregulate mRNA levels in a sequence-specific manner. By exploiting this biological process, investigators can interrogate and characterize gene behavior in cultured cells. While individually targeting genes is quite powerful, the production of libraries of siRNA or dsRNA targeting entire genomes allows for a specific cellular function to be assessed genome-wide in a matter of 1 week. When paired with high-throughput acquisition technology, genome-wide RNAi screening (GWS) has almost limitless potential to garner insights into key proteins and factors that regulate various biochemical processes.

Yongsheng Shi (ed.), *mRNA Processing: Methods and Protocols*, Methods in Molecular Biology, vol. 1648, DOI 10.1007/978-1-4939-7204-3_17, © Springer Science+Business Media LLC 2017

There are currently RNAi libraries available for several species [1, 2] that can be used on multiple platforms. Depending on the species chosen for the screen, the method of RNAi will vary significantly. To date, GWSs in whole organisms are conducive only to *Caenorhabditis elegans* [3] which not only exhibit classical RNAi (posttranscriptional gene silencing) but are also capable of transcriptional gene silencing (systemic RNAi), with the latter being heritable. *C. elegans* can be fed bacteria containing double-stranded RNA (dsRNA), soaked in a solution of dsRNA or injected with dsRNA [4–6]. For GWSs in other multicellular systems (including mammalian and *Drosophila melanogaster*), cell lines are used. The choice of the model system is predicated by the biological question being asked. Subsequently, further considerations are the technical challenge of how to achieve RNAi in the cells as well as cost. Factors that influence the cost of the assay include medium to maintain the cells, media supplements, transfection reagents, and screening facilities fees. Overall, the ease of delivery of RNAi and the robustness of the screen assay are the most critical considerations and are expanded upon below.

The most important difference between mammalian and *Drosophila melanogaster* GWS is the method of si/dsRNA delivery. Mammalian cells are either transfected with siRNA/shRNA (alternatively shRNA are delivered by lentivirus) while in the *Drosophila melanogaster* S2 cells, long dsRNA is delivered by "bathing" or transfection. In the "bathing method," fly cells are incubated with medium containing long dsRNA [7]. Long dsRNAs of roughly 500–800 bps are absorbed by the cells and are the substrate used to facilitate RNA interference, as the organism contains the machinery to process the long dsRNA—which covers a larger range of the target gene body and can more efficiently knock down the gene [8, 9]. Below, we describe the process of executing a reporter-based high-throughput RNAi screen in *D. melanogaster.*

2 Materials

- RNAi library.

Premade Library

Drosophila Screening Center (previously www.flyrnai.org now http://fgr.hms.harvard.edu) has unbiased genome-wide libraries and clustered sublibraries for screening.

The current DRSC Genome-wide RNAi Library (DRSC 2) contains ~13,900 genes.

Sheffield RNAi Screening Facility (http://www.rnai.group.shef.ac.uk/) has unbiased genome-wide library containing ~15,000 genes (protein and nonprotein coding).

cDNA Library Sources

Berkeley Drosophila Genome Project http://www.fruitfly.org/about/materials/cDNA-list.html.

Drosophila Genome Resource Center https://dgrc.bio.indiana.edu/clones/Catalog.

- Schneider's Drosophila Line 2 [D. Mel. (2), SL2] (ATCC® CRL1963™) (*see* **Note 1**).
- 25 °C humidified incubator.
- Reporter Gene (*see* **Note 2**).
- Effectene Transfection Reagent (Qiagen).
- Serum-Free Insect Medium (*see* **Note 1**).
- 100× Antimycotic/Antibiotic (Penicillin, Streptomycin and Amphotericin B).
- Blasticidin (10 mg/mL).
- Fetal bovine Serum (FBS).
- Multi-well plates 96- or 384-well plates for high-throughput screen, 6- and 24-well plates for the generation of stable lines (*see* **Note 3**).

3 Methods

3.1 Initial Considerations

A genome-wide RNAi screen requires the use of a comprehensive library to investigate the pathway of choice. There are premade libraries available from the screening facilities, Drosophila Screening Center and Sheffield RNAi Screening Facility. Alternatively, libraries can be created using the available cDNA libraries (*see* Subheading 2). The process of making user-generated libraries is described elsewhere [7]. Before embarking on any GWS it is critical to carefully think through how the screen will work and conduct several important pilot experiments. We have provided a simple schematic to guide in this process (Fig. 1, top). The first point to consider is whether the biological question is amenable to GWS. The best assays involve simple and scoreable readouts that use fluorescence or luminescence from a reporter or score an obvious morphological change using microscopy. The preferred design of a reporter system is one that will invoke a "gain-in-function" phenotype when a desired gene is scored/knocked down (discussed below). This helps in limiting the number of false positives when conducting the screen. Additionally, it is almost impossible to conduct a GWS without the use of a positive and negative control dsRNA in order to establish a dynamic range of detection and signal-to-noise ratios. Subsequently, once a positive control is obtained, we have found it useful to conduct a "blind screen" where positive controls are placed in multiple wells of a 96- or 384-well plate while

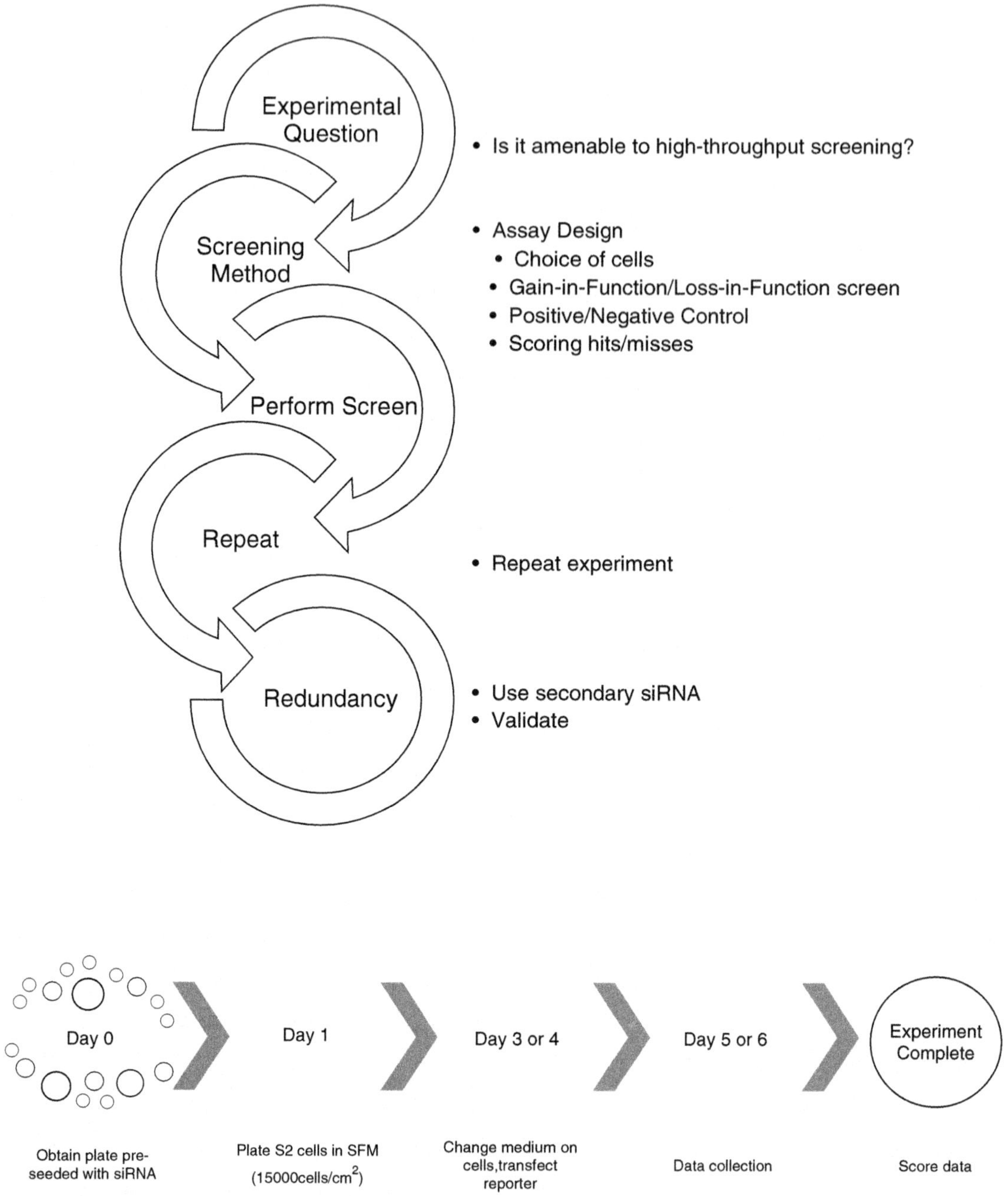

Fig. 1 The process of a high-throughput RNAi screen. (*Top*) Overview of the high-throughput RNAi screen from the rational and experimental design to the collection of data for analysis. (*Bottom*) Timeline a single round of the high-throughput RNAi screen, from the initial plating of the cells to data collection

the remainder of the wells are filled with negative control dsRNA. A screen is then performed with this test plate to determine how accurate the detector is at identifying positive hits. Finally, once the GWS is performed, if possible, a replicate GWS should be conducted in order to confirm positive scoring hits. After hits are attained, secondary screening should be performed using a second distinct dsRNA and finally, a rescue experiment using an RNAi-resistant cDNA should be conducted to rule against off-target effects.

3.2 Screening Method

There are several screening methods amenable to high-throughput screening. These include reporter-based screens, antibody-based screens, microscopic or fluorescence-activated cell sorting-based screens [10]. The reporter-based screens rely on ectopic expression of an enzymatically active protein such as luciferase, or the production of a fluorescent or chemiluminescent protein. An overview of various reporter genes is available elsewhere [11]. The expression of the reporter gene should be faithfully influenced by the biological question investigated [12]. This allows for the design of either a gain-of-function assay, which allows for screening for positive events rather than screening for negative events. We have provided a specific example of the concept of a gain-in-function reporter that has successfully been used in a GWS (Fig. 2 and ref. 13). In this specific case, a GFP cDNA is cloned downstream of the U7snRNA gene, thus giving rise to negligible GFP expression if transfected into untreated fly cells. If the members of the Integrator Complex 9 [14] are depleted using RNAi, then transcriptional readthrough should ensue giving rise to GFP expression. This is a clear and scoreable phenotype when assessed by fluorescence microscopy, Western blot analysis, or by high-throughput microscopy (*see* **Note 4**) (Fig. 2b–d).

In the loss-of-function reporter screens, expression of the reporter gene is decreased in response to depletion of the regulatory gene. The weakness of this approach is that reporter signal could diminish as the fitness of the cells diminishes in response to depletion of an essential gene. This will result in significantly more false positives. This can be mitigated if a normalizing signal is introduced, such as Hoechst stain or a constitutive reporter that is distinct from the scoring reporter.

3.3 Perform Screen

GWS Derivative #1: Transient Reporter Expression (for timeline and work flow, see Fig. 1 bottom)

1. Plate D.mel2 cells in either 96-well (1.2×10^5 cells/well) or 384-well ($3–4 \times 10^4$ cells/well) plate format containing pre-aliquoted and dried dsRNA library (three replicates per dsRNA if using your own custom-made dsRNA library). Include in each plate, positive control dsRNA and negative control dsRNA. Allow cells to incubate at 24 °C in a humidified incubator for 3 days. Cells are plated in the well in serum-free medium (SFM) with antibiotic/antimycotic.
2. On day 3, replenish SFM on cells. On day 4, transfect D.mel2 cells with predetermined amount of reporter (*see* **Note 2**)

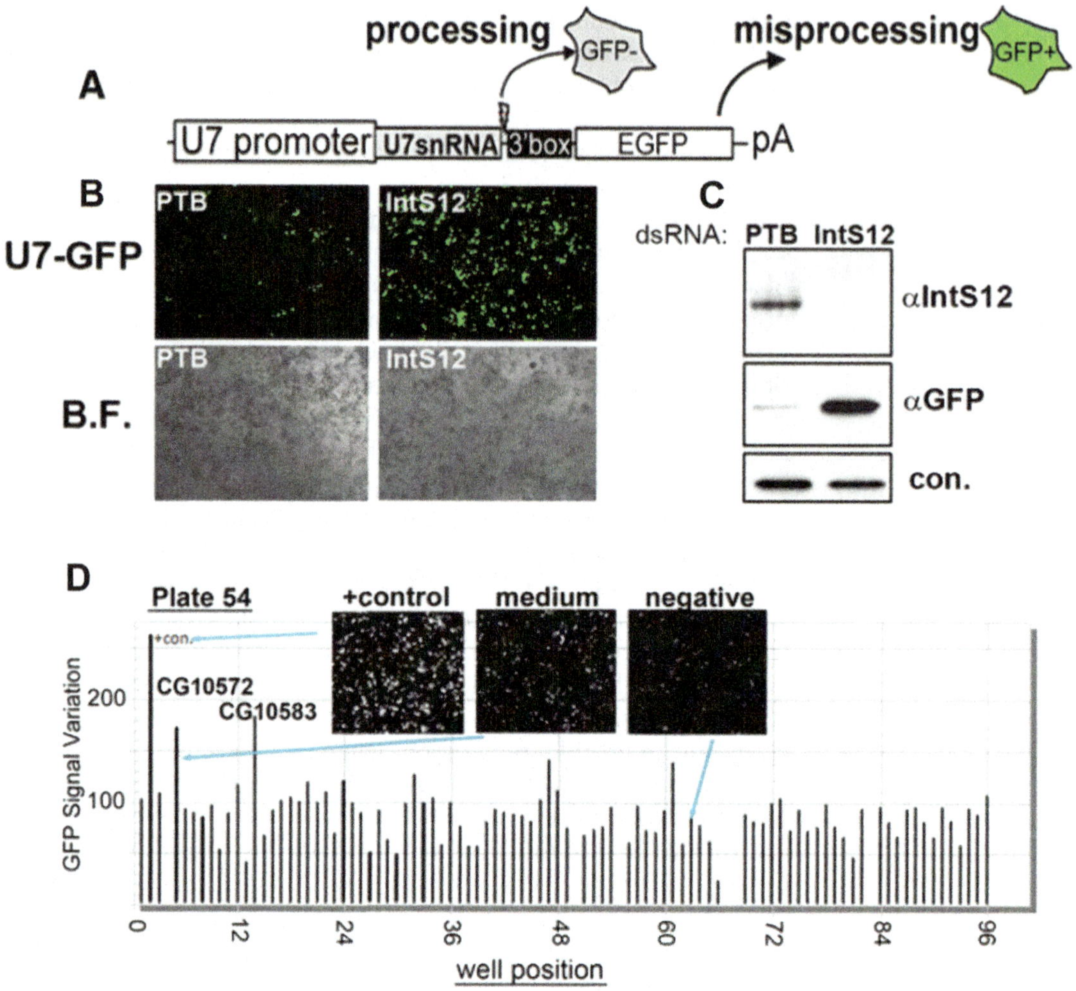

Fig. 2 Representation of the data acquired from a high-through put RNAi screen using a gain-of-function reporter. (**a**) Schematic of the gain-in-function reporter. (**b**) Fluorescence images of S2 cells transfected with the reporter after dsRNA treatment. (**c**) Western blot analysis of lysates from panel (**b**). (**d**) Bar graph quantification of fluorescent intensities from high-throughput screen in which positive and negative controls are re-identified with measurable and quantifiable differences. (Adapted from ref. 13)

using Effectene Reagent. Transfection with Effectene Reagent is done according to manufacturer's recommendation.

3. Allow cells to grow another 2 days (48 h) post reporter transfection before proceeding to data collection (*see* **Note 5**).
4. Data may be collected via microscopic quantification of fluorescence intensity from live cells or using a fluorescence plate reader.

GWS Derivative #2: Stable Cell Line Generation

1. On the day of transfection, D.mel2 cells (routinely cultured in SFM supplemented with antibiotic/antimycotic) are seeded in

a 24-well plate at a density of 5 × 10^5 cells/well. Cells are allowed to re-attach (*see* **Note 1**).

2. Mix 800 ng of reporter expression plasmid with 2.4 μL of enhancer reagent in 100 μL of EC buffer and incubate at room temperature for 5 min. If the expression plasmid does not contain the desired selection marker, a selection plasmid may be co-transfected in a ratio of 10:1 in favor of the expression plasmid (*see* **Note 6**). It is also essential to have an additional well of untransfected or mock-transfected S2 cells in order to ensure stable lines are created.
3. After 5 min has elapsed, add 4 μL of Effectene and gently mix. Incubate the mixture for another 10 min at room temperature, after which add dropwise to the cells.
4. On the next day, transfer the cells from the 24-well plate to a 6-well plate, and let the cells regrow overnight (*see* **Note 1**). The following day (48 h after transfection), change the medium on cells. Replenish cells 2 mL of Drosophila Schneider medium containing 10% FBS and the desired selection antibiotic (*see* **Note 6**).
5. Check the cells every day to monitor cell death and clonal expansion. In the event of significant cell death, change the medium every other day with great caution to avoid detaching cells from the plate.
6. Once the plate is over 50% confluent, and no more cell death is observed, passage the cells to a larger dish and return to serum-free medium conditions while maintaining the antibiotic selection.
7. Once the cells are growing exponentially again, test for the expression of the reporter by conducting a positive control RNAi experiment.
8. Plate stably transfected D.mel2 cells as described in **step 1** of *Transient Reporter Transfection and RNAi Screen*. NOTE: During this step, cells are not plated in the assay format in the presence of the selection antibiotic.
9. Allow cells to bathe in the presence of the dsRNA for 3 days transfection before proceeding to data collection (*see* **Note 5**).
10. Data may be collected via microscopic quantification of fluorescence intensity from live cells or fluorescence plate reader.

3.4 Screen and Hit Validation

- To test the validity and success of the screen, it is essential that both positive and ne.gative controls behave consistently and as expected within the parameters of the assay. As an example, in Fig. 2, we show data from a sample plate in an assay screening for regulators of snRNA 3′ end. Here, we show that the depletion of the positive control (Integrator subunit 12) results in an increase in the GFP intensity and was scored in the quantification [13].

- The hit validation can be done by applying the "two R's" methodology [15], wherein the R refers to redundancy and rescue (described below). Reduction in the scoring of false positives and identification off-target effects (OTE) can be accomplished through redundancy and rescue experiments.

Redundancy

- OTE are a significant concern when using *Drosophila melanogaster* for GWS because of the use of large dsRNA. When these dsRNA are processed into siRNA, they will have potential to target other genes even if mismatches are present. In addition, simple repeat sequences such as CAN (where N is any nucleotide) present in dsRNA libraries have been shown to increase OTE [16]. While the current second generation libraries are designed to minimize OTE [17], this does not negate the necessity for hit validation. In Fig. 3, we show an example of data validation [18], reproduced in a smaller scale. The dsRNA

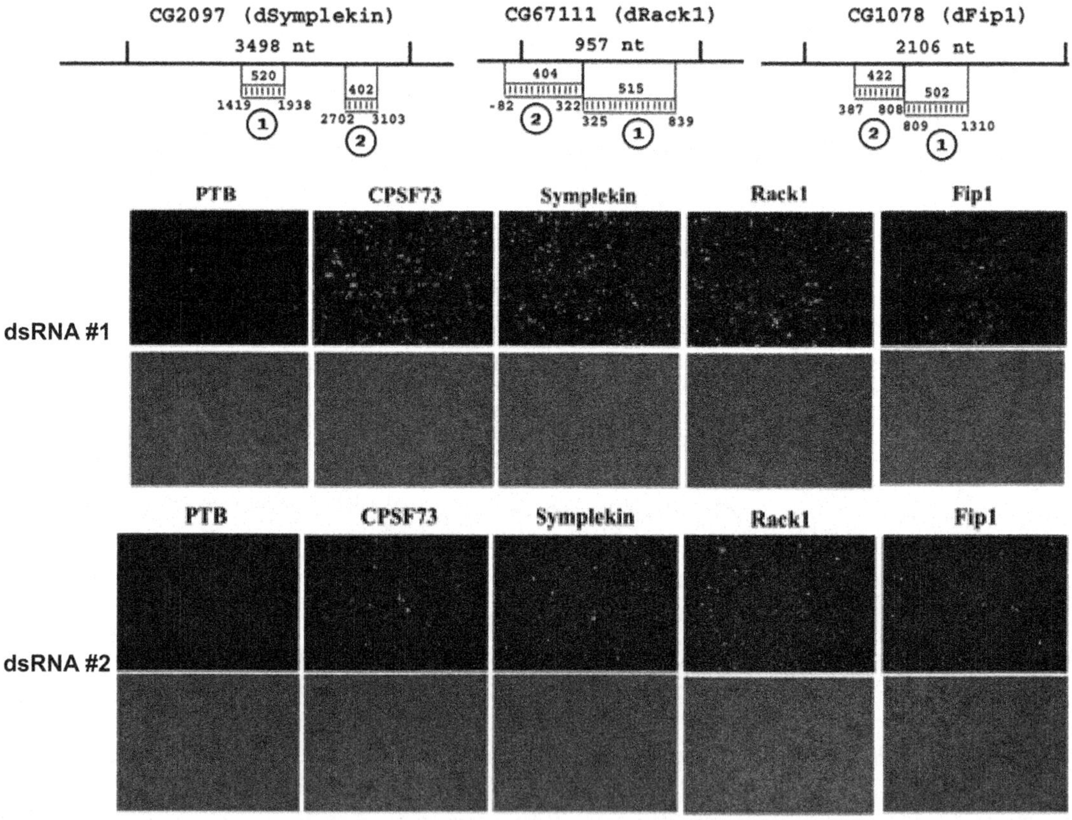

Fig. 3 Fluorescence images from validation of high-throughput screen using scaled down RNAi experiment with dsRNA from initial screen and novel nonoverlapping dsRNA. (*Top*) Schematic of dsRNA from initial screen and second user-designed dsRNA. (*Bottom*) Positive control is CPSF73 depletion and negative control is PTB depletion. Symplekin, Fip1, and Rack1 depletion result in expression of GFP in using dsRNA from initial high-throughput screen (*left*) and second dsRNA (*right*). (Adapted from ref. 18, used with permission from Elsevier)

from the initial screen that scored in the screen for factors regulating histone mRNA 3′ processing were used in repeated experiments, with the addition of a second user-designed non-overlapping dsRNA. In doing so, we can achieve both technical and biological validation (*see* **Note 7**).

Rescue

- The goal of a rescue experiment is to reverse the reporter phenotype associated with depletion of a positive hit dsRNA through the rescue of an RNAi-resistant cDNA to restore expression. Conducting RNAi rescue in S2 cells is not as straightforward as mammalian cells because the dsRNA used to deplete the mRNA is typically ~500 nucleotides in length. Regardless, this can be accomplished in one of three possible ways. The first is to create a dsRNA that targets the 3′UTR of the endogenous target followed by transfection of a plasmid that encodes the rescue cDNA that will not have its native 3′UTR present. The second method would be to express a cDNA from a distinct Drosophila species (i.e., *D. pseudoobscura*) that has enough changes to render it RNAi-resistant [19]. Finally, we have created synthetic fly genes using DNA synthesis services where the entire cDNA is re-created but altered to contain wobble base mutations throughout the dsRNA targeting regions [20].

4 Notes

1. D.mel2 cells, also known as S2 cells or SL2 cells, are derived from the embryo of *Drosophila melanogaster*. We use the D.mel2 cell line because they are adapted to grow in serum-free conditions. These cells are particularly amenable to GWS because the absence of serum reduces the risk of RNAse activity. Other cell lines derived from this parent line are also adapted to grow in serum-free conditions [7]. We have successfully used Sf-900 II SFM (Thermo) and SFX Insect (Hyclone) to culture our D.mel2 cells. D.mel2 cells can be cultured to grow loosely adherent or in suspension. We culture D.mel2 such that they adhere and have observed that the D.mel2 cells typically begin adhering to the plates typically within 1 h of plating, but this may vary empirically. D.mel2 cells will form a monolayer of cells that can easily be disturbed. To passage the cells, cells can be resuspended by gentle pipetting to ensure all the cells have detached. Also, when changing the medium on the cells care must be taken to minimize cell loss.
2. We recommend establishing a stock of reporter plasmid prior to experiment. The amount of reporter RNA to be transfected should be predetermined using a pilot experiment with known

positive control and negative control. Use the minimum amount of reporter, which gives a reproducible and significant (relative to the background and non-reporter transfected cells) scored using the chosen analysis method.

3. Multi-well plates used in the assay should be optimal for the assay method chosen. For example, clear and flat-bottom multiwall plates are appropriate for fluorescence detection [7].
4. We have used the transcriptional read-through reporters wherein the open reading frame (ORF) of green fluorescence protein (GFP) is placed downstream of the 3′ end processing event, in particular processing of the U7 snRNA, Histone mRNA [13, 18]. We have also described a modification of this reporter to assay 3′ end processing of polyadenylated mRNA in mammalian cells [21]. We chose to use a fluorescence reporter because it allows for live cell imaging and data collection, in lieu of a luciferase-based reporter that typically requires cell lysis and injection-enabled plate readers. We chose our fluorescent protein based on the fluorescence intensity over background and signal duration. A limitation of the reporter-based high-throughput screening performed independently of a dedicated screening facility is that using either fluorescence-based or luminescence-based reporters requires machinery and software capable of sensitive/quantitative analysis.
5. The length of the RNAi experiment may affect the degree of knockdown and impair resolution of primary and secondary effects [16].
6. We typically use Blasticidin (Invitrogen) for selection and stable line generation. If our expression plasmid lacks the marker for Blasticidin resistance, we typically co-transfect the expression plasmid with pCoBlast (Thermo).
7. We find the resource www.flybase.org an excellent tool to design and validate user-designed siRNAs. Moreover, the repository has information about homologues of fly proteins in other species, which can further assist with experimental redundancy if more than one model system can be tested to demonstrate similar phenotypes.

References

1. Echeverri CJ, Perrimon N (2006) High-throughput RNAi screening in cultured cells: a user's guide. Nat Rev Genet 7:373–384
2. Kuttenkeuler D, Boutros M (2004) Genome-wide RNAi as a route to gene function in Drosophila. Brief Funct Genomic Proteomic 3:168–176
3. Crotty S, Pipkin ME (2015) In vivo RNAi screens: concepts and applications. Trends Immunol 36:315–322
4. Maeda I, Kohara Y, Yamamoto M et al (2001) Large-scale analysis of gene function in *Caenorhabditis elegans* by high-throughput RNAi. Curr Biol 11:171–176

5. Kamath RS, Ahringer J (2003) Genome-wide RNAi screening in *Caenorhabditis elegans*. Methods (San Diego, Calif) 30:313–321
6. Gönczy P, Echeverri C, Oegema K et al (2000) Functional genomic analysis of cell division in *C. elegans* using RNAi of genes on chromosome III. Nature 408:331–336
7. Ramadan N, Flockhart I, Booker M et al (2007) Design and implementation of high-throughput RNAi screens in cultured Drosophila cells. Nat Protoc 2:2245–2264
8. Shih JD, Fitzgerald MC, Sutherlin M et al (2009) The SID-1 double-stranded RNA transporter is not selective for dsRNA length. RNA 15:384–390
9. Meister G, Tuschl T (2004) Mechanisms of gene silencing by double-stranded RNA. Nature 431:343–349
10. Perrimon N, Mathey-Prevot B (2007) Applications of high-throughput RNA interference screens to problems in cell and developmental biology. Genetics 175:7–16
11. Schenborn E, Groskreutz D (1999) Reporter gene vectors and assays. Mol Biotechnol 13:29–44
12. Grimm S (2004) The art and design of genetic screens: mammalian culture cells. Nat Rev Genet 5:179–189
13. Chen J, Ezzeddine N, Waltenspiel B et al (2012) An RNAi screen identifies additional members of the Drosophila integrator complex and a requirement for cyclin C/Cdk8 in snRNA 3′-end formation. RNA (New York, NY) 18:2148–2156
14. Baillat D, Wagner EJ (2015) Integrator: surprisingly diverse functions in gene expression. Trends Biochem Sci 40:257–264
15. Echeverri CJ, Beachy PA, Baum B et al (2006) Minimizing the risk of reporting false positives in large-scale RNAi screens. Nat Methods 3:777–779
16. Ma Y, Creanga A, Lum L et al (2006) Prevalence of off-target effects in Drosophila RNA interference screens. Nature 443:359–363
17. Fisher KH, Wright VM, Taylor A et al (2012) Advances in genome-wide RNAi cellular screens: a case study using the Drosophila JAK/STAT pathway. BMC Genomics 13:506
18. Wagner EJ, Burch BD, Godfrey AC et al (2007) A genome-wide RNA interference screen reveals that variant histones are necessary for replication-dependent histone pre-mRNA processing. Mol Cell 28:692–699
19. Kondo S, Booker M, Perrimon N (2009) Cross-species RNAi rescue platform in *Drosophila melanogaster*. Genetics 183:1165–1173
20. Chen J, Waltenspiel B, Warren WD et al (2013) Functional analysis of the integrator subunit 12 identifies a microdomain that mediates activation of the Drosophila integrator complex. J Biol Chem 288:4867–4877
21. Peart N, Wagner EJ (2016) Gain-of-function reporters for analysis of mRNA 3′end formation: design and optimization. BioTechniques 60:137–140

Chapter 18

Assaying RNA Structure Inside Living Cells with SHAPE

Chao Feng, Dalen Chan, and Robert C. Spitale

Abstract

RNA molecules have emerged as key players in nearly every facet of gene regulation. Such functions are governed by RNA's unique ability to fold into intricate secondary and tertiary structures. In order to understand, on the molecular level, how such structures from several chemical and enzymatic methods have been developed. One such method is RNA Selective Hydroxyl Acylation analyzed by Primer Extension, or SHAPE. SHAPE has emerged as a premier method of measuring RNA structure, with recent developments extending SHAPE into living cells. Here, we describe the use of SHAPE for measuring RNA structure inside living cells.

Key words RNA structure, SHAPE, Chemical Probing

1 Introduction

Investigating molecules as they function in a living environment remains one of the foremost challenges in molecular biology. Once regarded as a passive genetic transporter, RNA is now recognized as a critical regulator of essential biological processes, including cell growth, differentiation, and cell death [1–3]. Many functions of RNA—including binding of protein partners, splicing, editing, translation, and processing for silencing—are intimately coupled to the formation, breaking, or switching of RNA secondary and tertiary structure [4–6]. Cancer and viral infectivity are among the numerous disease states that have been associated with specific mRNA structures [7–11]. Without structural analyses, investigations of RNAs give incomplete pictures of the accessible conformational ensemble; such descriptions are necessary to fully understand their modes of action while providing a basis to test or improve therapies designed to sense, disrupt, or activate specific functional RNAs. Therefore, insight into the role that structure

Chao Feng and Dalen Chan contributed equally to this work.

Yongsheng Shi (ed.), *mRNA Processing: Methods and Protocols*, Methods in Molecular Biology, vol. 1648,
DOI 10.1007/978-1-4939-7204-3_18, © Springer Science+Business Media LLC 2017

plays in controlling posttranscriptional pathways could be a powerful inroad into better understanding RNA's roles in biology and medicine.

Chemical mapping can be used to read out RNA structure. Direct methylation of the RNA through reaction with dimethylsulfate, hydroxyl radical cleavage of solvent exposed residues, hydroxyl acylation by solution electrophiles, or by metal catalyzed cleavage are currently the most rigorous methods for determining the secondary and tertiary structure of RNAs [12]. While these methods are quite useful for traditional probing techniques, many of these chemical methods are not generalizable and often have biases in their reactivities; this limits the ability to obtain structural information for every nucleotide.

A general chemical functionality that all RNAs have is a 2′-hydroxyl group in the ribose ring. 2′-hydroxyl is more chemically reactive at single-stranded positions than at nucleotides constrained by base pairing [13, 14]. Reagents that modify RNA as a function of RNA 2′ hydroxyl reactivity can be used to read out RNA structure by reverse-transcription primer extension in experiments termed RNA SHAPE (Selective 2′ Hydroxyl Acylation and Primer Extension) [13, 14]. SHAPE has been proven to be highly accurate at identifying single-stranded residues and can also be merged with structure prediction programs to obtain highly accurate secondary structure models [15]. SHAPE is widely used and has been proven to be very powerful at measuring RNA structure in vitro and has found widespread utility for identifying and mapping RNA-protein interactions.

We recently reported the development, synthesis, and utilization of the first SHAPE reagents to measure RNA structure inside cells (Fig. 1) [16]. Such reagents can permeate into the nucleus of mammalian cells and read out specific RNA structural arrangements that are unique to the structure of RNA in its native environment.

Within this Methods chapter we detail the protocol for measuring RNA structure inside living cells using the nicotinoylimidazole reagent, NAI. NAI can permeate into mammalian, yeast, bacteria, and fly cells and read out specific RNA structural arrangements that are unique to the structure of RNA in its native environment. NAI has already been utilized by many labs to measure RNA structure by SHAPE, both inside and outside of cells, and as such serves as a highly suitable reagent for the comparison of RNA structures [17–19].

2 Materials

All solutions should be prepared using ultrapure water treated with RNase inhibitor DEPC (Diethylpyrocarbonate). Prepare and store all reagents at room temperature, unless otherwise noted. Diligently

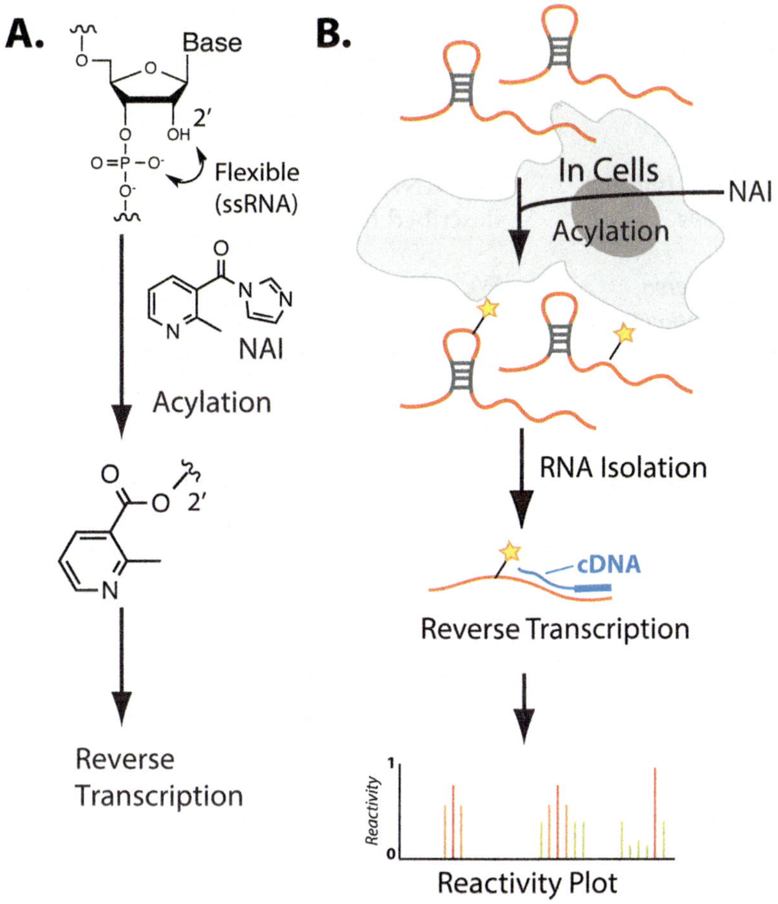

Fig. 1 Schematic of RNA acylation and SHAPE. (**a**) Chemical reaction for SHAPE with NAI. (**b**) Schematic of the in-cell SHAPE experiment

follow all waste disposal recommendations as well as MSDS protocols for handling.

2.1 Materials for Primer Kinase

Radioactive ATP can be purchased at Perkin Elmer. CAUTION: When working with radioactivity, take appropriate precautions to avoid contamination of the experimenter and the surroundings. Carry out the experiment and dispose of wastes in an appropriately designated area, following the guidelines provided by the local radiation safety officer. T4 polynucleotide kinase can be purchased at New England Biolabs.

2.2 Reagents for the Preparation of NAI

2-methylnicotinic acid and 1,1′-carbonyldiimidazole were used as received. Anhydrous dimethylsulfoxide (DMSO) was used as received.

2.3 Media for Cell Culture

Depending on the cell type media its supplements can be vastly different. Cells utilized in the lab are purchased from American Type Culture Collection (ATCC) and prepared according to their best practices (https://www.atcc.org/en.aspx).

2.4 RNA Isolation from Cells

For RNA isolation from mammalian cells we follow protocols described for the TriZol reagent.

2.5 Reverse Transcription Reaction and Buffer

We utilize the reverse transcriptase SuperScript III. SSIII engineered for higher thermostability, with a longer half-life at 50 °C, and reduced RNase H activity, resulting in higher yields of full-length cDNA for more complete gene product representation.

2.6 Denaturing PAGE Gels (RT Analysis)

10% acrylamide/Bis solution (29.2:0.8) acrylamide:bis: Weigh 10 g of acrylamide monomer and 0.8 g Bis (cross-linker) and transfer to a 100 mL graduated cylinder containing about 40 mL of water. Add a spatula of AG 501-X8 (D) mixed-resin beads and mix for about 30 min. Make up to 100 mL with water and filter through a 0.45 μm Corning filter (*see* **Note 3**). Store at 4 °C, in a bottle wrapped with aluminum foil. Obtain an ammonium persulfate 10% solution in water. N,N,N,N′-Tetramethyl-ethylenediamine (TEMED) is used to initiate polymerization. TEMED should be stored at 4 °C.

2.7 Gel Electrophoresis

Sequencing apparatus for high-resolution polyacrylamide gel electrophoresis. We recommend using a vertical denaturing gel (the dimensions depend on the length of the gel and cDNA that you are running). We focus on ^{32}P imaging and as such a Phosphorimaging instrument and screen will be necessary.

3 Methods

The methods outlined herein are based on utilizing the 5SrRNA as a positive control. 5SrRNA is a robust positive control because it is highly expressed, it is expressed in every cell line, and a high-resolution crystal structure is available for direct comparison of SHAPE reactivities. We also are focused on mammalian cell culturing in this Methods chapter; however, in-cell SHAPE with NAI has been demonstrated to work in yeast, bacteria, and fly cells and as such the methods post-lysis and RNA isolation can be applied to all other species and RNA structure probing.

3.1 Kinase of 5SrRNA Reverse Transcription Primer

1. Mix 2 pmoles primer with 1 μL 10× kinase buffer [0.5 M Tris–HCl, pH 7.5, 0.1 M $MgCl_2$, 50 mM dithiothreitol, 10 mM EDTA], 4 μL (40 μCi, 3000 Ci/mmole) g-[32P]-rATP, and enough water to bring the volume to 9 μL.

2. Add 10 units (1 μL) T4 polynucleotide kinase and incubate at 37 °C for 30–45 min.
3. Stop by adding 90 μL of 10 mM Tris–HCl pH 8.0, 1 mM EDTA and heating to 65 °C for 5–10 min.
4. Note: the primer sequence for the positive control 5SrRNA is: 5′-AAAGCCTACAGCACCCGGTAT-3′(REF).

3.2 Isolation of the Primer Perform Denaturing PAGE

1. There are some nice references for a detailed description of setting up denaturing gel electrophoresis [20, 21].
2. Use bromophenol blue and xylene cyanol. 20% denaturing gel solution works well for the isolation of primers in the rage of 15–20 nucleotides. Use bromophenol blue and xylene cyanol to mark the gel. The primer will run between the two dyes.
3. For the isolation of the primer, remove the gel from the rig and place down on Whatman paper. Then wrap the gel in plastic wrap and then place inside a phosphorimaging screen for 1 min.
4. Image the phosphorimaging screen and align the image with the gel to mark the site of the band.
5. Cut and remove the band into a 1.5 mL eppendorf tube. Crush the gel with a 1 mL pipette tip. Add 400 μL of gel soaking buffer (Crush Soak Buffer (CSB)—200 mM NaCl, 10 mM Tris–HCl (pH 7.5), 1 mM EDTA (pH 8) using MilliQ water/ NanoPure water. And let it incubate at 50 °C overnight.
6. The next day add the crushed solution to Costar Spin-X Centrifuge Tube (Sigma Aldrich) and follow the manufacturer's protocol for isolation and precipitation of nucleic acids.
7. Resuspend radiolabeled primer and measure ^{32}P counts by scintillation.

3.3 Preparation of Cell SHAPE Reagent, NAI

1. A schematic of the chemical reaction is shown in Fig. 2a.
2. NAI stocks are routinely made at 1 M in DMSO. This is to have 0.1 M final concentration in RNA structure probing experiments.
3. First, 137 mg (1 mmol) 2-methylnicotinic acid was dissolved in 0.5 mL anhydrous DMSO. A solution of 162 mg (1 mmol) 1,1′-carbonyldiimidazole in 0.5 mL anhydrous DMSO was added dropwise over 5 min. The resulting solution is stirred at room temperature until gas evolution was complete and then stirred at room temperature for 1 h. The resulting solution is used as a 1.0 M stock solution (assuming complete conversion) containing a 1:1 mixture of the desired compound and imidazole.
4. The solution was frozen at −80 °C when not in use. The reagent is stable for several months if stored in anhydrous DMSO at −80 °C.

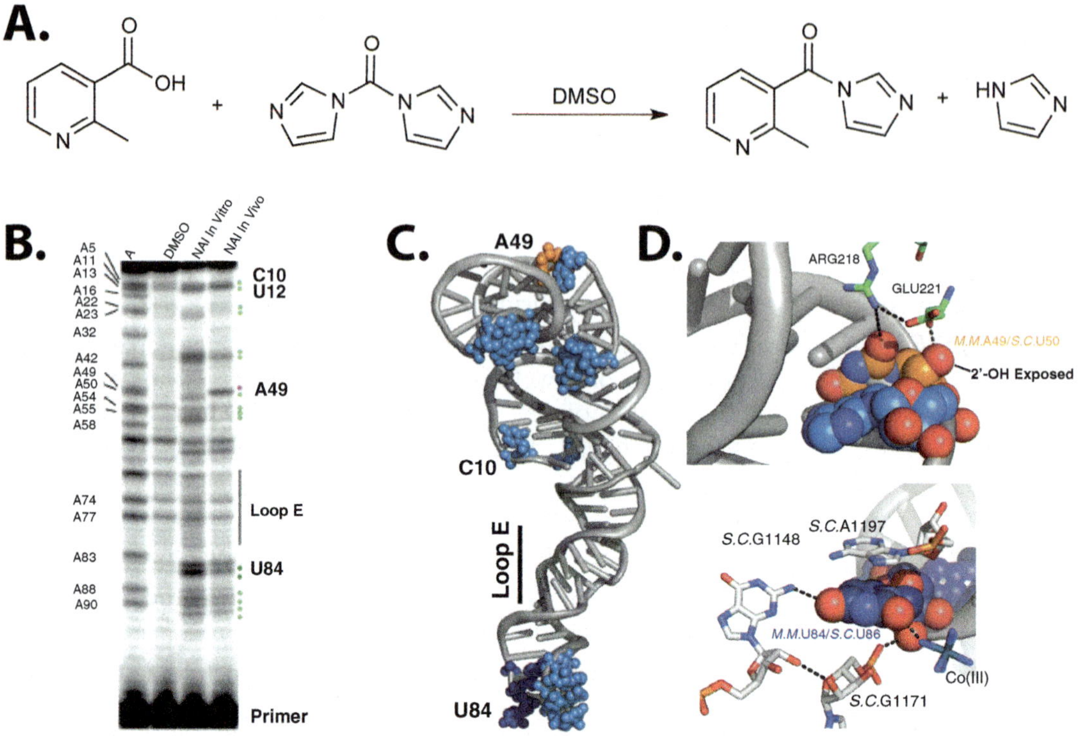

Fig. 2 Preparation of NAI and Expected Results for SHAPE on 5SrRNA inside cells. (**a**) Chemical reaction to prepare NAI. (**b**) Typical results for 5SrRNA SHAPE probing. (**c**) Difference SHAPE map for 5SrRNA SHAPE probing. Residues in *blue* are less reactive in vivo. *Orange* is more reactive in vivo. (**d**) Close-up structural analysis of 5SrRNA difference SHAPE reactivity

5. The stock solution should be warmed to room temperature before opening.

3.4 Cell Culture and Maintenance (Mammalian Cells)

1. Grow adherent or suspension cells of choice under standard conditions to ~70–80% confluency. Several published protocols are available for mammalian cell culturing [22].
2. Remove cells from the incubator.
3. Decant media into the appropriate waste container.
4. Wash cells three times with 1× PBS buffer, pH 7.2. Add an additional 10 mL solution of X PBS to the cells.
5. Gently scrape cells and decant them into a 10 mL Falcon tube. Spin down the cells at 67 × *g* (1000 rpm) for 2 min.
6. Decant the PBS media from the spin down cell pellet. Add 450 μL of 1XPBS and GENTLY mix cells up and down by pipette.
7. To the PBS-cell solution add 50 μL of the 1 M NAI solution in DMSO. Allow the cells to rock gently at 37 °C for 15 min. *See* **Note 1**.

3.5 Isolation of Total RNA from Cells

1. After incubation spin down the cell suspension at 268 × *g* (2000 rpm) for 1 min.
2. Decant the NAI solution and dispose in the appropriate waste container.
3. To the cells at 5 mL of Trizol. Follow the Trizol procedure for RNA isolation and precipitation.

3.6 Reverse Transcription of Modified RNA

1. The typical samples to be reverse transcribed, in parallel, consist of: (1) DMSO negative control, (2) NAI Experimental, (3) dideoxynucleotide sequencing samples (preferably at least two sequencing samples to compare for sequencing the RNA, *see* **Note 3**).
2. Set up the primer annealing reactions in a total volume of 7 μL.
3. To up to one total RNA from cells, add SSIII buffer and the kinased RT primer and set aside on ice until all RT reactions are mixed.
4. Anneal the primer at 95 °C for 2 min and then add to ice and let site for 10 min. *See* **Note 2**.
5. To the mixture add the dNTP mix and DTT (both provided with the SSIII master mix).
6. Heat the mixture to 50 °C for 1 min.
7. Add the SSIII enzyme to the mix and allow it to incubate at 50 °C for 15 min. *See* **Note 4**.
8. Add the SSIII buffer and the dNTP mix.
9. To the mix add the kinased RT primer and set aside on ice until all RT reactions are mixed.
10. To stop the reactions add one-tenth volume of 2 M sodium hydroxide (to cleave the RNA from the RNA-cDNA hybride). Incubate at 95 °C for an additional 5 min.
11. Add an equal volume of denaturing Gel Loading Buffer II and incubate at 95 °C for an additional 2 min. Place the samples directly on ice. At this point the samples can be stored at −80 °C for several months or directly loaded onto the denaturing PAGE gel.

3.7 Denaturing Gel Electrophoresis

For a detailed description of setting up denaturing gel electrophoresis, please refer to the following guide [20]. Typical results, which should be in agreement with the published result for 5S, are shown in Fig. 2b.

3.8 Analysis of Reverse Transcription Stops

For the analysis of reverse transcription stops and mapping to the sequence of the RNA of interest, we use the program Semi-Automated Footprinting Analysis, or SAFA. A SAFA download and instructions are available in references in the literature and on the web here: https://simtk.org/projects/safa/. As a positive control, the 5SrRNA data is shown in Fig. 2. Reactivities can then be mapped back to the 5S rRNA crystal structure to identify sites that are different in vivo and in vitro. *See* **Note 5**.

4 Notes

1. The time and concentration of NAI will need to be optimized for the cell type and RNA that is being probed. We have found that 15 min at 100 mM final concentration is highly suitable for typical structure probing reactions. For other cells, such as bacteria and yeast, which have tougher membrane environments, the concentration may need to be higher. We have demonstrated that NAI is highly soluble, up to even 500 mM final in buffer.
2. The addition of ddNTPs for sequencing lanes will need to be optimized. Too much ddNTPs will give truncated cDNA stops near the 3′-end. Too little will give weak signal in the gel.
3. The annealing temperature is best determined empirically but can be approximated by adding four times the number of GC base pairs and two times the number of AT and AU base pairs in the hybrid, and subtracting 5. Bringing the mixture to the annealing temperature and allowing it to slow cool at least 15 degrees over the course of 45 min usually ensures that efficient and specific annealing will occur. Allowing the temperature to go too low (e.g., placing the annealing reactions on ice) may encourage the formation of hybrids between the oligonucleotide and nontarget RNAs. This will need to be optimized for robust extension.
4. The extension time for SSIII is typically 15 min for most RNAs. However, longer extensions of those with RNAs that are predicted to have a significant amount of structure due to high GC content may need longer extensions. This will need to be optimized for robust extension.
5. Interpreting the cDNA pattern on the gel is by far the most important part. Often times, RNase contamination can lead to spurious RT stops or fragmentation that can alter the signal-to-noise ratio of the experiment. If there is little modification of the RNA increase the concentration of NAI by two-fold. If the cDNA stops are near the primer and there are weak cDNA stops near the top of the gel, the RNA was over-modified; decrease the concentration by two-fold. The parameters for NAI modification will need to be optimized to have effective signal-to-noise ratios and clean cDNA gels that can be interpreted.

Acknowledgments

RNA research in the Spitale lab is supported by startup funds from the University of California, Irvine, the NIH Director's New Innovator Award (1DP2GM119164 RCS) and 1RO1MH109588 (RCS). Chao Feng and Dalen Chan contributed equally to this manuscript.

References

1. Amaral PP, Mattick JS (2008) Noncoding RNA in development. Mamm Genome 19(7–8):454–492. doi:10.1007/s00335-008-9136-7
2. Qureshi IA, Mattick JS, Mehler MF (2010) Long non-coding RNAs in nervous system function and disease. Brain Res 1338:20–35. doi:10.1016/j.brainres.2010.03.110
3. Pang KC, Frith MC, Mattick JS (2006) Rapid evolution of noncoding RNAs: lack of conservation does not mean lack of function. Trends Genet 22(1):1–5. doi:10.1016/j.tig.2005.10.003
4. Mercer TR, Mattick JS (2013) Structure and function of long noncoding RNAs in epigenetic regulation. Nat Struct Mol Biol 20(3):300–307. doi:10.1038/nsmb.2480
5. Luo Z, Yang Q, Yang L (2016) RNA structure switches RBP binding. Mol Cell 64(2):219–220. doi:10.1016/j.molcel.2016.10.006
6. Bevilacqua PC, Ritchey LE, Su Z, Assmann SM (2016) Genome-wide analysis of RNA secondary structure. Annu Rev Genet. doi:10.1146/annurev-genet-120215-035034
7. Sabarinathan R, Wenzel A, Novotny P, Tang X, Kalari KR, Gorodkin J (2014) Transcriptome-wide analysis of UTRs in non-small cell lung cancer reveals cancer-related genes with SNV-induced changes on RNA secondary structure and miRNA target sites. PLoS One 9(1):e82699. doi:10.1371/journal.pone.0082699
8. Lv L, Xiao XY, ZH G, Zeng FQ, Huang LQ, Jiang GS (2011) Silencing USP22 by asymmetric structure of interfering RNA inhibits proliferation and induces cell cycle arrest in bladder cancer cells. Mol Cell Biochem 346(1–2):11–21. doi:10.1007/s11010-010-0585-4
9. Kwon SJ, Kim KH (2006) The SL1 stem-loop structure at the 5′-end of potato virus X RNA is required for efficient binding to host proteins and for viral infectivity. Mol Cells 21(1):63–75
10. Ohno T, Okada Y, Shimotohno K, Miura K, Shinshi H (1976) Enzymatic removal of the 5′-terminal methylated blocked structure of tobacco mosaic virus RNA and its effects on infectivity and reconstitution with coat protein. FEBS Lett 67(2):209–213
11. Black DR, Connell CJ, Merigan TC (1973) Structure and infectivity of picornaviral RNA encapsidated by cowpea chlorotic mottle virus protein. J Virol 12(6):1209–1215
12. Kubota M, Tran C, Spitale RC (2015) Progress and challenges for chemical probing of RNA structure inside living cells. Nat Chem Biol 11(12):933–941. doi:10.1038/nchembio.1958
13. Wilkinson KA, Merino EJ, Weeks KM (2005) RNA SHAPE chemistry reveals nonhierarchical interactions dominate equilibrium structural transitions in tRNA(Asp) transcripts. J Am Chem Soc 127(13):4659–4667. doi:10.1021/ja0436749
14. Merino EJ, Wilkinson KA, Coughlan JL, Weeks KM (2005) RNA structure analysis at single nucleotide resolution by selective 2′-hydroxyl acylation and primer extension (SHAPE). J Am Chem Soc 127(12):4223–4231. doi:10.1021/ja043822v
15. Deigan KE, Li TW, Mathews DH, Weeks KM (2009) Accurate SHAPE-directed RNA structure determination. Proc Natl Acad Sci U S A 106(1):97–102. doi:10.1073/pnas.0806929106
16. Spitale RC, Crisalli P, Flynn RA, Torre EA, Kool ET, Chang HY (2013) RNA SHAPE analysis in living cells. Nat Chem Biol 9(1):18–20. doi:10.1038/nchembio.1131
17. Pirakitikulr N, Kohlway A, Lindenbach BD, Pyle AM (2016) The coding region of the HCV genome contains a network of regulatory RNA structures. Mol Cell 62(1):111–120. doi:10.1016/j.molcel.2016.01.024
18. Hector RD, Burlacu E, Aitken S, Le Bihan T, Tuijtel M, Zaplatina A, Cook AG, Granneman S (2014) Snapshots of pre-rRNA structural flexibility reveal eukaryotic 40S assembly dynamics at nucleotide resolution. Nucleic Acids Res 42(19):12138–12154. doi:10.1093/nar/gku815

19. Guo JU, Bartel DP (2016) RNA G-quadruplexes are globally unfolded in eukaryotic cells and depleted in bacteria. Science 353(6306). doi:10.1126/science.aaf5371
20. Summer H, Gramer R, Droge P (2009) Denaturing urea polyacrylamide gel electrophoresis (urea PAGE). J Vis Exp 32. doi:10.3791/1485
21. Slatko BE, Albright LM (2001) Denaturing gel electrophoresis for sequencing. In: Frederick MA et al (eds) Current protocols in molecular biology. John Wiley and Sons, Inc., New Jersey. Chapter 7:Unit7 6. doi:10.1002/0471142727.mb0706s16
22. Phelan K, May KM (2016) Basic techniques in mammalian cell tissue culture. Curr Protoc Toxicol 70:A 3B 1–A 3B 22. doi:10.1002/cptx.13

Erratum to: Robust, Cost-Effective Profiling of RNA Binding Protein Targets with Single-end Enhanced Crosslinking and Immunoprecipitation (seCLIP)

Eric L. Van Nostrand, Thai B. Nguyen, Chelsea Gelboin-Burkhart, Ruth Wang, Steven M. Blue, Gabriel A. Pratt, Ashley L. Louie, and Gene W. Yeo

Erratum to:
Chapter 14 in: Yongsheng Shi (ed.), *mRNA Processing: Methods and Protocols*, Methods in Molecular Biology, vol. 1648, https://doi.org/10.1007/978-1-4939-7204-3_14

The chapter 'Robust, Cost-Effective Profiling of RNA Binding Protein Targets with Single-end Enhanced Crosslinking and Immunoprecipitation (seCLIP)' was originally published © Springer Science+Business Media LLC, but has now been made available © The Author(s) under a CC BY 4.0 license.

The updated online version of this chapter can be found at https://doi.org/10.1007/978-1-4939-7204-3_14

Yongsheng Shi (ed.), *mRNA Processing: Methods and Protocols*, Methods in Molecular Biology, vol. 1648, https://doi.org/10.1007/978-1-4939-7204-3_19, © Springer Science+Business Media LLC 2017

INDEX

Yongsheng Shi (ed.), *mRNA Processing: Methods and Protocols*, Methods in Molecular Biology, vol. 1648,
DOI 10.1007/978-1-4939-7204-3, © Springer Science+Business Media LLC 2017

CPSIA information can be obtained
at www.ICGtesting.com
Printed in the USA
BVOW07*0412050318
509703BV00002B/6/P

9 781493 972036